AF352924

Hybrid Systems for Marine Energy Harvesting

Hybrid Systems for Marine Energy Harvesting

Editors

Paulo Jorge Rosa-Santos
Francisco Taveira Pinto
Mario López Gallego
Claudio Alexis Rodríguez Castillo

MDPI • Basel • Beijing • Wuhan • Barcelona • Belgrade • Manchester • Tokyo • Cluj • Tianjin

Editors
Paulo Jorge Rosa-Santos
Faculty of Engineering of the
Universidade do Porto
Portugal

Francisco Taveira Pinto
Faculdade de Engenharia da
Universidade do Porto (FEUP)
Portugal

Mario López Gallego
University of Oviedo
Spain

Claudio Alexis Rodríguez Castillo
Federal University of Rio de
Janeiro
Brazil

Editorial Office
MDPI
St. Alban-Anlage 66
4052 Basel, Switzerland

This is a reprint of articles from the Special Issue published online in the open access journal *Journal of Marine Science and Engineering* (ISSN 2077-1312) (available at: https://www.mdpi.com/journal/jmse/special_issues/marine_energy_harvesting).

For citation purposes, cite each article independently as indicated on the article page online and as indicated below:

LastName, A.A.; LastName, B.B.; LastName, C.C. Article Title. *Journal Name* **Year**, *Volume Number*, Page Range.

ISBN 978-3-0365-4627-8 (Hbk)
ISBN 978-3-0365-4628-5 (PDF)

Cover image courtesy of Paulo Rosa Santos

Contents

About the Editors . . . . **vii**

Paulo Rosa-Santos, Francisco Taveira-Pinto, Mario López and Claudio A. Rodríguez
Hybrid Systems for Marine Energy Harvesting
Reprinted from: *J. Mar. Sci. Eng.* **2022**, *10*, 633, doi:10.3390/jmse10050633 . . . . **1**

Theofano I. Koutrouveli, Enrico Di Lauro, Luciana das Neves, Tomás Calheiros-Cabral,
Paulo Rosa-Santos and Francisco Taveira-Pinto
Proof of Concept of a Breakwater-Integrated Hybrid Wave Energy Converter Using a
Composite Modelling Approach
Reprinted from: *J. Mar. Sci. Eng.* **2021**, *9*, 226, doi:10.3390/jmse902022 . . . . **5**

Daniel Clemente, Tomás Calheiros-Cabral, Paulo Rosa-Santos and Francisco Taveira-Pinto
Hydraulic and Structural Assessment of a Rubble-Mound Breakwater with a Hybrid Wave
Energy Converter
Reprinted from: *J. Mar. Sci. Eng.* **2021**, *9*, 922, doi:10.3390/jmse9090922 . . . . **33**

Dimitrios N. Konispoliatis, Spyridon A. Mavrakos and Georgios M. Katsaounis
Theoretical Evaluation of the Hydrodynamic Characteristics of Arrays of Vertical Axisymmetric
Floaters of Arbitrary Shape in front of a Vertical Breakwater
Reprinted from: *J. Mar. Sci. Eng.* **2020**, *8*, 62, doi:10.3390/jmse8010062 . . . . **51**

Gianmaria Giannini, Irina Temiz, Paulo Rosa-Santos, Zahra Shahroozi, Victor Ramos,
Malin Göteman, Jens Engström, Sandy Day and Francisco Taveira-Pinto
Wave Energy Converter Power Take-Off System Scaling and Physical Modelling
Reprinted from: *J. Mar. Sci. Eng.* **2020**, *8*, 632, doi:10.3390/jmse8090632 . . . . **89**

Mario López, Noel Rodríguez and Gregorio Iglesias
Combined Floating Offshore Wind and Solar PV
Reprinted from: *J. Mar. Sci. Eng.* **2020**, *8*, 576, doi:10.3390/jmse8080576 . . . . **111**

Victor Ramos, Gianmaria Giannini, Tomás Calheiros-Cabral, Mario López,
Paulo Rosa-Santos and Francisco Taveira-Pinto
Assessing the Effectiveness of a Novel WEC Concept as a Co-Located Solution for
Offshore Wind Farms
Reprinted from: *J. Mar. Sci. Eng.* **2022**, *10*, 267, doi:10.3390/jmse10020267 . . . . **131**

Shufang Qin, Jun Fan, Haiming Zhang, Junwei Su and Yi Wang
Flume Experiments on Energy Conversion Behavior for Oscillating Buoy Devices Interacting
with Different Wave Types
Reprinted from: *J. Mar. Sci. Eng.* **2021**, *9*, 852, doi:10.3390/jmse9080852 . . . . **153**

About the Editors

Paulo Jorge Rosa-Santos

Paulo Jorge Rosa-Santos is Associate Professor with Habilitation at FEUP—Faculty of Engineering of the University of Porto, Portugal, and Integrated Member of CIIMAR—Interdisciplinary Centre of Marine and Environmental Research of the University of Porto. He completed his PhD in Civil Engineering (2010) entitled *Behaviour of Moored Ships in Harbours* and obtained his Habilitation/Aggregation (2021) with his thesis *Harbour Breakwaters and Marine Renewable Energy*. He is currently co-coordinator of the Marine Energy group of CIIMAR and responsible for the subunit "Marine Renewable Energy". His main research interests are in the application of physical and numerical modelling in the study of offshore, port, and coastal-related topics, namely marine renewable energy, wave energy, breakwaters, port terminals, berthing structures, climate change impacts, coastal protection. He is a Chartered Engineer (2004), Senior Member (2013), and Specialist in Hydraulics and Water Resources (2019) of the Portuguese Engineering Council (Ordem dos Engenheiros). He has served as a member of the Board of the Portuguese Water Resources Association—North Branch since 2016 and participated as a researcher and/or PI in more than 18 national and international R&D projects. In addition, he has collaborated in more than 50 industry projects and consultancy activities on coastal, port, and ocean engineering, coastal management, applied hydraulics, and on the development of technologies for wave energy conversion.

Francisco Taveira Pinto

Francisco Taveira Pinto is Full Professor of Hydraulics, Water Resources and Environment, and Coastal and Port Engineering (2010) at FEUP—Faculty of Engineering of the University of Porto, Portugal, Deputy Director of the Civil Engineering Department, President of the Hydraulics and Water Resources Institute (IHRH), and Director of the Hydraulics Laboratory of FEUP. He obtained his degree in Civil Engineering (1989), PhD for *Analysis of Oscillations and Velocity Fields in the Proximity of Submerged Breakwaters under the Wave Action* (2002), and Aggregate (2007) from the University of Porto. He is an Integrated Member of CIIMAR—the Interdisciplinary Centre of Marine and Environmental Research of the University of Porto and Coordinator of NEMAR—Hydraulic Structures and Marine Energy Group of FEUP and of the Marine Energy group of CIIMAR, focused on the numerical and physical modelling of wave energy converters and coastal and port structures. He was designated a Specialist in Hydraulics and Water Resources (Coastal and Port Engineering) by the Order of Engineers and Vice-President of the General Assembly of the Portuguese Water Partnership. He was previously member of the executive committee of EUCC—The Coastal Union Mediterranean Centre, Barcelona, Spain, President of the Board of the General Assembly of the Portuguese Water Resources Association (APRH) and Coordinator of the Specialization in Hydraulics and Water Resources of the Order of Engineers. He was also Chairman of the Maritime Hydraulics Committee of the International Association for Hydro-Environment Engineering and Research (IAHR). He has expertise in the fields of marine renewable energy, coastal, and port engineering. He has been involved in physical modelling works for more than 30 years and participated in several EU projects.

Mario López Gallego

Mario López Gallego is Associate Professor in the Dept. of Construction and Manufacturing Engineering and member of the Dynamics, Materials and Structures research group (DyMAST) of the University of Oviedo (Spain). His main area of interest and expertise is the numerical assessment of coastal and offshore structures such as floating breakwaters, renewable technology, and aquaculture farms. He completed his PhD in Engineering with the thesis Long waves and their effects on vessels at the port of Ferrol at the University of Santiago de Compostela (2014). In 2015, he was granted by the Post-Doctoral Program of the Foundation for Science and Technology of Portugal for the development of a novel wave energy conversion system in the Faculty of Engineering of the University of Porto. Since 2016, he has worked at the University of Oviedo, where he started building a research team in offshore technologies. He has participated in more than 20 public and private research projects and authored 25 journal papers in JCR-indexed journals. Since 2019, he is the principal investigator in his institution of PORTOS, an European project to promote and develop renewable energy in ports. He also lectures several courses on continuum mechanics and structures at both BSc and MSc level in the Polytechnic School of Mieres, where he is also the Vice-Dean for Quality.

Claudio Alexis Rodríguez Castillo

Claudio Alexis Rodríguez Castillo is Associate Professor in the Department of Naval Architecture and Ocean Engineering and researcher at the Laboratory of Ocean Technology (LaOceano) of the Federal University of Rio de Janeiro (Brazil). His main area of interest and expertise is the numerical and experimental assessment of the stability and dynamics of ships and offshore structures such as oil and gas platforms and ocean renewable devices. He completed his PhD in Ocean Engineering focusing on ship hydrodynamics in the Federal University of Rio de Janeiro, Brazil (2010). During 2017–2018, he was awarded with a Post-Doctoral research grant from FCT—Fundação para a Ciência e a Tecnologia—Portugal, within the framework of Project "OPWEC—Optimizing Wave Energy Converters" developed at the Faculty of Engineering of the University of Porto. Since 2014, has actively participated in the International Towing Tank Conference in the specialist committees of Ocean Engineering and Stability in Waves. He has authored more than 30 scientific journal papers, several of them dealing with the hydrodynamics of energy conversion devices. He is currently working as a research associate of the University of Strathclyde (UK) in the Ocean Renewable Energy Fuel (Ocean REFuel) Project.

Journal of *Marine Science and Engineering*

MDPI

Editorial

Hybrid Systems for Marine Energy Harvesting

Paulo Rosa-Santos [1,2,*], Francisco Taveira-Pinto [1,2], Mario López [3] and Claudio A. Rodríguez [4]

[1] Hydraulics, Water Resources and Environment Division, Department of Civil Engineering, Faculty of Engineering of the University of Porto (FEUP), Rua Dr. Roberto Frias, s/n, 4200-465 Porto, Portugal; fpinto@fe.up.pt

[2] Interdisciplinary Centre of Marine and Environmental Research of the University of Porto (CIIMAR), Terminal de Cruzeiros do Porto de Leixões, Av. General Norton de Matos, s/n, 4450-208 Matosinhos, Portugal

[3] DyMAST Research Group, Department of Construction and Manufacturing Engineering, University of Oviedo, EPM, C/Gonzalo Gutiérrez Quirós s/n, 33600 Mieres, Asturias, Spain; mario.lopez@uniovi.es

[4] Laboratory of Ocean Technology (LabOceano), Department of Naval Architecture and Ocean Engineering, Federal University of Rio de Janeiro, Rua Leopoldo de Meis 569—Prédio 1, Cidade Universitária, Rio de Janeiro CEP 21941-855, Brazil; claudiorc@oceanica.ufrj.br

* Correspondence: pjrsantos@fe.up.pt

Citation: Rosa-Santos, P.; Taveira-Pinto, F.; López, M.; Rodríguez, C.A. Hybrid Systems for Marine Energy Harvesting. *J. Mar. Sci. Eng.* **2022**, *10*, 633. https://doi.org/10.3390/jmse10050633

Received: 8 March 2022
Accepted: 27 April 2022
Published: 6 May 2022

Publisher's Note: MDPI stays neutral with regard to jurisdictional claims in published maps and institutional affiliations.

The marine renewable energy (MRE) industry is being stimulated by the growing world energy demand, climate change mitigation policies, and land-use conflicts. The ocean is our largest primary global renewable energy resource. Moreover, it offers large free areas to harvest its potential. However, the harsh and variable marine environment challenges the survivability, resilience and reliability of current harvesting technologies and, consequently, impacts the cost and the affordability of electricity generation. Exhaustive research is being conducted to reduce the levelized cost of energy (LCoE) of MRE by reducing cost drivers and increasing generation. In this context, the use of hybrid systems appears to be a promising solution.

Wind and waves are concentrated forms of solar energy, comprising vast and underexploited offshore resources with different degrees of technological maturity. A large number of wave energy converters (WECs) have been proposed—from the pioneer concept of Salter to the innovative concept CECO—but a fully commercial WEC has not been reached yet. Offshore wind turbines (OWTs) fixed to the seabed in shallow waters have been generating electricity for decades. Harvesting the full potential of offshore wind requires moving to deeper waters, where reliable floating foundations are required. Similar to WECs, although a wide variety of floating OWT technologies have been proposed, no particular kind stands out above the others yet.

Another promising MRE is offshore floating photovoltaics (FPV). The deployment of FPV systems in continental bodies of freshwater has grown exponentially in recent years, and is still unexploited in many world regions with a high potential. In the meantime, the transition of FPV technology to the marine environment has already started, and different ad hoc solutions are being developed.

In addition to the challenge of developing cost-effective technologies (affordability), MRE faces key issues in contributing to the decarbonization (sustainability) of the energy sector: its intermittent nature (reliability), and location/marine environmental constraints (resilience). These aspects result in an inadequate energy generation, i.e., the ability of an existing generation portfolio to match power demand at all times. In fact, as renewable shares increase, scheduling power generation becomes more challenging. Conventional operational and technical solutions to integrate higher penetrations include forecasting, demand response, flexible generation, larger balancing areas, balancing area cooperation, and fast scheduling and dispatch, among others. Hybridization also contributes to mitigate the excessive production costs and the intermittent nature of MRE.

In the scope of this Special Issue, hybridization should be understood in three different manners:

(i) The combination of technologies to harvest different MREs (e.g., WECs integrated into OWTs);

(ii) The combination of different working principles to harvest the same resource (e.g., oscillating water columns with overtopping devices to harvest wave energy);

(iii) The integration of harvesting technologies in multifunctional platforms and structures (e.g., the integration of WECs in breakwaters, oil and gas platforms, or aquaculture platforms);

(iv) The co-location of different harvesting technologies and multi-resource systems, for example, combining wind turbines, tidal turbines, diesel generators, and pumped hydroelectric storage.

Hybridization also fosters synergies among the different MRE resources and harvesting technologies. Possible synergies might be related to opportunities for cost reduction in the technology itself—including deployment, operation, maintenance, and decommissioning costs—but also to improvements in the quantity and quality of energy generation. Technological synergies include:

- Common grid infrastructure. Access to the power grid is assured when a MRE technology is deployed in an existing coastal infrastructure, which reduces deployment costs. In cases of hybrid conversion technologies, these costs would be shared;
- Shared logistics during the lifetime of the project;
- Common substructure, mooring or foundation systems. The mooring system can represent about 10% of the capital expenditure for a WEC and may reach higher values for an OWT. In cases of integration with other coastal structures, such as breakwaters, this may be shared. Furthermore, breakwaters offer a reliable foundation for WECs and OWTs. Therefore, shared foundation systems would significantly reduce the total deployment costs;
- Shadow effects. A wave energy farm may significantly reduce the energy of incident sea states, and therefore dampen wave forces on other structures. Combining WECs with other MRE technologies in the same farm and with an appropriate layout may reduce structure requirements, as well as the corresponding manufacturing, operational and maintenance (O&M) costs.

As for the generation synergies of hybridization, these include:

- Enhanced energy yield. Relative to a traditional MRE farm, a hybrid farm would increase the total energy production per unit surface area. For example, if OWTs are combined with offshore FPVs, the production per surface area can increase by a factor of ten;
- Better predictability. Wave resources are easier to predict and more constant in time than wind resources. Therefore, the combination of both will reduce the system balancing costs;
- Smoothed power output. Co-located technologies have been proven as a solution to smooth the power output and reduce the disadvantages of MRE when penetrating the energy mix. Some examples include combined wind–wave farms, wind–solar farms and wave–solar farms.
- In the same weather systems, wave climate peaks trail behind wind peaks. Consequently, a combined exploitation will result in a reduction in sudden disconnections from the electric grid, an increase in availability (reducing the number of hours of non-activity), and a more accurate output forecast.

The minimization of negative impacts of MRE applications on biodiversity and ecosystems is considered an essential condition for the environmental permission of such projects and for social acceptance. Hybrid installations are expected to cause less environmental impacts compared to independent installations. Since, in some areas of MRE production, industrial fishing operations or intense navigation may not be allowed, these areas may potentially become shelters of marine life or artificial reefs.

This Special Issue presents a set of works on hybrid systems for marine energy conversion which focus on innovative numerical and/or experimental research, and demonstrate projects on the development of technologies to harness MREs. Seven articles have been compiled addressing the following hybrid systems:

- The integration of WECs in port breakwaters;
- The combination of offshore wind and solar-power harvesting technologies;
- The combination of WECs and offshore wind turbines.

As evidenced in this Special Issue, at present, there is significant interest in the integration of WECs into port breakwaters or the areas nearby these structures. Given their exposure to ocean waves and the high energy consumption of port facilities, this is considered an excellent combination that could also increase the sustainability of seaports.

For instance, a novel hybrid WEC (HWEC), which combines an overtopping device with an oscillating water column system on a breakwater, is proposed and assessed for proof of concept in the first study of the Special Issue [1]. The study of this HWEC, which uses air and water turbines, is performed using a composite modelling approach, i.e., combining physical and CFD numerical modelling. The preliminary research findings demonstrate that the HWEC is viable in both rubble-mound and vertical breakwaters, without detrimental effects to their hydraulic performance and structural response/stability. A succeeding experiment assessment reports significant reductions in overtopping discharges in a case study of rubble-mound breakwater, which is considered a highly beneficial contribution of the hybrid concept to the functional performance of the structure. However, the authors recommended a careful analysis of its structural stability and damage potential, in order to maintain safety levels [2]. This study also points out that traditional damage assessment parameters should be applied with care when non-conventional structures are analysed, for example, rubble-mound breakwaters with integrated WECs.

Another study proposes taking advantage of the Bragg reflection induced by multiple submerged breakwaters (partial standing wave field) in nearshore areas, in combination with a fully standing wave field generated by wave reflection on a vertical wall, in order to enhance the performance of an oscillating buoy WEC [3]. The flume experiments conducted in the study demonstrate that the standing wave field created could enhance the energy conversion performance of the WEC. Furthermore, the Special Issue also presents a research study that deals with the numerical determination of hydrodynamic loads on arrays of floaters in front of a vertical breakwater [4]. In the study, different arrays and shapes of vertical axisymmetric floaters are numerically investigated. The image method is applied to simulate the effect of the breakwater on the array, and the multiple scattering approach is used to evaluate the interaction phenomena among the WECs. The geometrical characteristics of the floater had a greater influence on the values of its hydrodynamic forces and coefficients than both the existence of the breakwater and the arrangement of the array with respect to the incoming wave.

Concerning the combination of offshore wind and solar power, a case study from the north of Spain is presented [5]. The power resources and production are assessed based on high-resolution data and the technical specifications of commercial wind turbines and solar photovoltaic (PV) panels. It is reported that, relative to a typical offshore wind farm, a combined offshore wind–solar farm can increase the capacity and energy production per unit surface area by factors of ten and seven, respectively. Hence, the utilization of the marine space is optimized, and the power output is significantly smoother.

Regarding the combination of offshore wind energy devices and WECs, one of the papers assesses the performance of a combined farm located on the northern coast of Portugal, in terms of energy production, power smoothing, and LCoE [6]. The authors conclude that the co-located farm increased the annual energy production by approximately 19% in comparison with the stand-alone wind farm for the studied region. Furthermore, they report that the LCoE of the hybrid farm reduces drastically in comparison with the stand-alone wave farm, presenting a value of USD 0.116 per kWh. Finally, one of the articles discusses the challenges in the scaling and physical modelling of power take-off systems for WECs [7]. Set-up enhancements, calibration practices, and error estimation methods are covered. Recommendations on the organization and conduction of the experiments are also provided, together with a brief overview of three different case studies.

In summary, this Special Issue covers attractive current topics in MRE and its applications in the context of hybrid systems, providing valuable contributions and motivation for further developments.

Author Contributions: The authors jointly carried out the activities leading to the Special Issue and co-wrote this editorial. All authors have read and agreed to the published version of the manuscript.

Funding: This work was partially supported by the project PORTOS—Ports Towards Energy Self-Sufficiency, with the reference EAPA_784/2018 and co-financed by the Interreg Atlantic Area Program through the European Regional Development Fund, and by the OCEANERA-NET COFUND project WEC4Ports—A hybrid Wave Energy Converter for Ports, with the reference OCEANERA/0004/2019, under the frame of FCT. This work was funded by the project ATLANTIDA (NORTE-01-0145-FEDER-000040), supported by the North Portugal Regional Operational Programme (NORTE2020), under the PORTUGAL 2020 Partnership Agreement and through the European Regional Development Fund (ERDF).

Acknowledgments: The authors wish to thank all contributors to this Special Issue as well as the professional and efficient JMSE editorial staff without whose excellent assistance this issue would not have been possible. Additionally, the authors thank Cheryl Huo and Angela Xia for their assistance.

Conflicts of Interest: The authors declare no conflict of interest.

References

1. Koutrouveli, T.; Di Lauro, E.; das Neves, L.; Calheiros-Cabral, T.; Rosa-Santos, P.; Taveira-Pinto, F. Proof of Concept of a Breakwater-Integrated Hybrid Wave Energy Converter Using a Composite Modelling Approach. *J. Mar. Sci. Eng.* **2021**, *9*, 226. [CrossRef]
2. Clemente, D.; Calheiros-Cabral, T.; Rosa-Santos, P.; Taveira-Pinto, F. Hydraulic and Structural Assessment of a Rubble-Mound Breakwater with a Hybrid Wave Energy Converter. *J. Mar. Sci. Eng.* **2021**, *9*, 922. [CrossRef]
3. Qin, S.; Fan, J.; Zhang, H.; Su, J.; Wang, Y. Flume Experiments on Energy Conversion Behavior for Oscillating Buoy Devices Interacting with Different Wave Types. *J. Mar. Sci. Eng.* **2021**, *9*, 852. [CrossRef]
4. Konispoliatis, D.; Mavrakos, S.; Katsaounis, G. Theoretical Evaluation of the Hydrodynamic Characteristics of Arrays of Vertical Axisymmetric Floaters of Arbitrary Shape in front of a Vertical Breakwater. *J. Mar. Sci. Eng.* **2020**, *8*, 62. [CrossRef]
5. López, M.; Rodríguez, N.; Iglesias, G. Combined Floating Offshore Wind and Solar PV. *J. Mar. Sci. Eng.* **2020**, *8*, 576. [CrossRef]
6. Ramos, V.; Giannini, G.; Calheiros-Cabral, T.; López, M.; Rosa-Santos, P.; Taveira-Pinto, F. Assessing the Effectiveness of a Novel WEC Concept as a Co-Located Solution for Offshore Wind Farms. *J. Mar. Sci. Eng.* **2022**, *10*, 267. [CrossRef]
7. Giannini, G.; Temiz, I.; Rosa-Santos, P.; Shahroozi, Z.; Ramos, V.; Göteman, M.; Engström, J.; Day, S.; Taveira-Pinto, F. Wave Energy Converter Power Take-Off System Scaling and Physical Modelling. *J. Mar. Sci. Eng.* **2020**, *8*, 632. [CrossRef]

Journal of
*Marine Science
and Engineering*

Article

Proof of Concept of a Breakwater-Integrated Hybrid Wave Energy Converter Using a Composite Modelling Approach

Theofano I. Koutrouveli [1], Enrico Di Lauro [1], Luciana das Neves [1,2,*], Tomás Calheiros-Cabral [2],
Paulo Rosa-Santos [2] and Francisco Taveira-Pinto [2]

[1] IMDC—International Marine and Dredging Consultants, Van Immerseelstraat 66, 2018 Antwerp, Belgium; theofano.koutrouveli@imdc.be (T.I.K.); enrico.di.lauro@imdc.be (E.D.L.)

[2] Department of Civil Engineering, FEUP—Faculty of Engineering of the University of Porto, Rua Dr. Roberto Frias, s/n, 4200-465 Porto, Portugal; tcabral@fe.up.pt (T.C.-C.); pjrsantos@fe.up.pt (P.R.-S.); fpinto@fe.up.pt (F.T.-P.)

* Correspondence: luciana.das.neves@imdc.be

Abstract: Despite the efforts of developers, investors and scientific community, the successful development of a competitive wave energy industry is proving elusive. One of the most important barriers against wave energy conversion is the efficiency of the devices compared with all the associated costs over the lifetime of an electricity generating plant, which translates into a very high Levelised Cost of Energy (LCoE) compared to that of other renewable energy technologies such as wind or solar photovoltaic. Furthermore, the industrial roll-out of Wave Energy Converter (WEC) devices is severely hampered by problems related to their reliability and operability, particularly in open waters and during harsh environmental sea conditions. WEC technologies in multi-purpose breakwaters—i.e., a structure that retains its primary function of providing sheltered conditions for port operations to develop and includes electricity production as an added co-benefit—appears to be a promising approach to improve cost-effectiveness in terms of energy production. This paper presents the proof of concept study of a novel hybrid-WEC (HWEC) that uses two well understood power generating technologies, air and water turbines, integrated in breakwaters, by means of a composite modelling approach. Preliminary results indicate: firstly, hybridisation is an adequate approach to harness the available energy most efficiently over a wide range of metocean conditions; secondly, the hydraulic performance of the breakwater improves; finally, no evident negative impacts in the overall structural stability specific to the integration were observed.

Keywords: Computational Fluid Dynamics (CFD) modelling; physical model testing; Hybrid-Wave Energy Converter (HWEC); composite modelling approach; Oscillating Water Column (OWC); Overtopping Device (OTD); multi-purpose breakwater

Citation: Koutrouveli, T.I.; Di Lauro, E.; das Neves, L.; Calheiros-Cabral, T.; Rosa-Santos, P.; Taveira-Pinto, F. Proof of Concept of a Breakwater-Integrated Hybrid Wave Energy Converter Using a Composite Modelling Approach. *J. Mar. Sci. Eng.* **2021**, *9*, 226. https://doi.org/10.3390/jmse9020226

Academic Editor: Eva Loukogeorgaki

Received: 30 December 2020
Accepted: 16 February 2021
Published: 20 February 2021

1. Introduction

Over the past years, global decarbonisation efforts have accelerated the quest for finding diverse, clean and renewable energy sources in order to achieve the targets in energy and climate that have been set out. Oceans are a safe, inexhaustible and largely untapped source of renewable energy that may significantly contribute to the electrical energy supply of vast coastal regions in the world [1]. Therefore, is not surprising that more than a thousand Wave Energy Converters (hereafter WECs) have been designed and developed worldwide [2] to varying stages, from concept to pre-commercial roll out.

Despite the efforts of developers, investors and scientific community, the successful development of a competitive wave energy industry is proving elusive. One of the most important barriers against wave energy conversion is the efficiency of the devices compared with all the associated costs over the lifetime of an electricity generating plant, which translates into a very high LCoE compared to that of other renewable energy technologies such as wind or solar photovoltaic. High costs make the devices still not economically feasible and not competitive on the global market [3]. Furthermore, the industrial roll-out of WEC devices is severely hampered by problems related to their reliability and operability, particularly in open waters and during harsh environmental sea conditions. There are currently a vast number of concepts at varying stages of development, supported by a very substantial investment effort and the hope that a competitive wave energy industry will successfully develop soon; however, to date, only a few technologies have gone from the research design and prototyping stage through demonstration and pre-commercial phases.

Because wave energy is a sizeable prize, relentless and determined effort should continue to be put in order to find the solutions to overcome the similar challenges experienced by wave energy technologies while progressing through each development phase. In this respect, many authors (e.g., [4–6]) argue that a good idea is integrating WEC devices in harbour defence structures, in so-called multi-purpose breakwaters. Firstly, capital expenditure costs can be shared between WEC and breakwater [7]. Secondly, the breakwater offers a reliable foundation for the WEC [8]. Furthermore, access to the main power grid is assured [9]. Finally, access for maintenance and repairs is facilitated [10]. In such multi-purpose breakwaters, the structure retains its primary function—that is providing sheltered conditions for port operations to develop—and electricity production is an added co-benefit. Therefore, a part of the energy that is usually dissipated by traditional harbour defence structures and consequently lost, can be captured.

Despite the added advantages of the integration, the wave energy locally available to be extracted with nearshore technologies is lower than that available in deeper water conditions. Furthermore, the integration of a WEC is not adequate for all breakwater types, in particular for already existing breakwaters, as the benefit of the integration strongly depends on the configuration, geographic location and orientation of the breakwater with respect to the prevalent wave climate conditions. Studying the combination of a WEC and a breakwater requires considering distinct aspects that concern both, the multi-purpose breakwater as a harbour defence structure, and the renewable energy device. It is clear that the most important goal for the integration of a WEC device into harbour breakwaters is to guarantee their functionality as protection works. Detailed studies are required to evaluate the hydraulic performance and structural response/stability of the breakwater, in terms of wave reflection, wave overtopping, wave loading, as well as local and global stability. Secondly, the performance of the system as a WEC device, focusing on the evaluation of the WEC generated power and performance as well as the reliability and survivability of the WEC technology when subjected to harsh maritime conditions, need to be examined. The integration of WECs into breakwaters has shown significant results in terms of hydraulic response of the device [10]. However, more research and innovation is needed in terms of improving the energy produced by WEC devices. In this context, the hybridisation of different WEC technologies in multi-purpose breakwaters appears to be a promising approach to improve their efficiency. This is the underlying idea behind a novel and innovative device, the hybrid-WEC (HWEC), which was first presented under the framework of the OCEANERA-NET Second Joint Call 2016 project SE@PORTS project (e.g., [11,12]) and is now being further developed under the OCEANERA-NET Second Joint Call 2019 project WEC4PORTS. The principal goal of the SE@PORTS project was to assess the suitability of existing WECs to be integrated in port infrastructures, bringing the selected concepts to the next Technology Readiness Level (TRL). By combining the main current and well-established principles in harnessing wave energy (i.e., the Oscillating Water Column—OWC and the Overtopping Device—OTD), the HWEC concept is aiming to exploit the strengths of each technology and overcome their individual limitations when implemented separately, thereby presenting a breakthrough and efficient approach to harness the wave energy at ports.

Following this introductory section on the integration of WECs into harbour defence structures and the novel concept HWEC, the present paper is organised as follows: Section 2 is devoted to the description of the performances of the OWC integrated into a breakwater, while Section 3 describes the performances of the OTD device embedded in breakwaters;

the detailed description and discussion of the performances of the HWEC obtained with the physical and numerical model tests under the SE@PORTS project is presented in Section 4; finally, the main conclusions are drawn in the last section.

2. Breakwater-Integrated OWC

With two pre-commercial pilot plants currently installed in Spain [4] and Italy [5], the breakwater-integrated OWC is currently the most prominent and probably the most successful example of a multi-purpose breakwater for energy production. The technology consists of a breakwater with a chamber, having a submerged and open inlet in its exposed face to allow water to enter into the chamber (Figure 1). An air-duct with an air turbine installed connects the chamber to the atmosphere. Under wave loading, the air in the chamber is alternately compressed and decompressed, creating a bi-directional flow through the duct. This flow drives a self-rectifying turbine connected with a generator for electricity production.

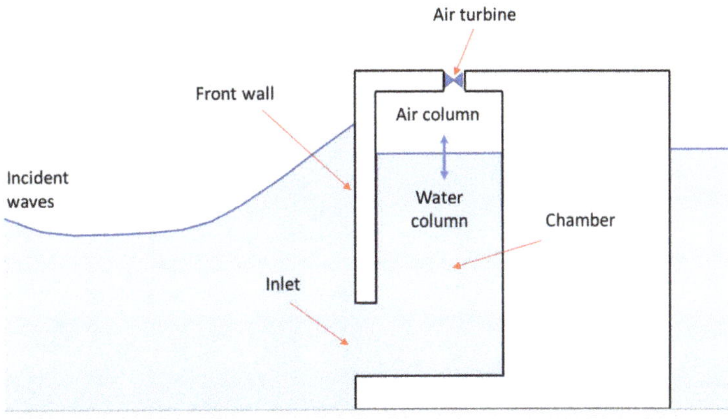

Figure 1. Scheme of the OWC integrated into a vertical caisson breakwater.

The main advantage of the breakwater-integrated OWC is the simplicity of the concept, a rigid structure without any moving parts except the rotor of the air turbine. In addition, air turbines are a very well-understood technology that has been around for many decades. The integration of an OWC in a new vertical caisson breakwater is probably the most straightforward solution of integration from the economic, construction and operation viewpoints [11]. With respect to disadvantages, the low operability of the OWC device in highly energetic sea states, caused by problems of loss of efficiency or damage of the turbine due to excessive air pressure in the chamber and green water droplets/jets reaching it, is the most challenging. Finally, it is worth underlining that the integration of this device in existing breakwaters might change the original cross-section of the breakwater, which can have a relevant impact on the structural response/stability to wave loading, as well as in its hydraulic performance mostly in terms of wave reflection and overtopping.

2.1. Structural Response/Stability to Waves

Although many studies using physical and numerical modelling have been conducted to evaluate the wave loading and the structural response to incident waves of Breakwater-integrated OWC, there is no widely accepted and well-established methodology to design such innovative configurations. Early studies conducted in Japan by Takahashi [13] suggested calculating the force distribution on a vertical breakwater integrated OWC adopting a modified Goda's formulae [14], which is traditionally adopted for conventional vertical caisson breakwaters. Other authors focused their research on the influence of the device geometry on the wave pressure distribution at the OWC front wall [15,16] and back wall inside the chamber [17], as well as on making a detailed evaluation of the maximum impact pressure [18,19]. New design methods for the calculation of the wave forces on an OWC front wall were presented by Thiruvenkatasamy et al. [20], Patterson [21], Huang et al. [22], Liu et al. [23] conducted physical model testing focusing on the analysis of the stability of a vertical breakwater-integrated OWC. Research results generally confirm that the wave forces on a vertical breakwater-integrated OWC are smaller than those acting on a conventional vertical breakwaters. Similar results confirming that Goda's formulae overestimates the maximum resultant forces for a vertical breakwater-integrated OWC, were obtained by Kuo et al. [24] based on measured data obtained in physical model testing of a vertical breakwater-integrated OWC under regular waves. Conversely, according to Kuo et al. [24], Goda's formulae underestimates the momentum due to a different wave-induced pressure distribution on the surface of the structure, which could affect the overall stability of the OWC caissons against overturning. Viviano et al. [25] evaluated the wave loading on these innovative breakwaters focusing on the influence of the turbine-orifice opening, while Naty et al. [26] showed that the wave forces on the front wall for an optimised OWC configuration could be accurately predicted by Sainflou [27] as used for vertical walls if they were increased by a safety factor equal to 1.1. Ashlin et al. [28] compared measured horizontal wave forces with those obtained by applying the Goda's formulae [14] to show that this formula overpredicts the shoreward peak forces by 46–90%, while underpredicting the seaward peaks by 5–50%. More recently, Pawitan et al. [29] proposed a novel method to estimate forces acting on an OWC chamber in a vertical caisson breakwater, validated against results from large scale physical model measurements described in Allsop et al. [30].

Despite a few discrepancies caused by different OWC configurations across the research conducted over the last decades on vertical breakwater-integrated OWC, results suggest that the nature and magnitude of wave loading acting on the front-wall of a OWC device differs from those acting on conventional vertical breakwaters. In particular, the total horizontal forces can be greater or lower than those acting on conventional vertical breakwaters, depending on the inner geometry of the chamber and turbine dimensions. Research on structural stability/response of a breakwater-integrated OWC was carried out essentially for vertical breakwaters to date. Very little or no studies focusing on a rubble-mound breakwater-integrated OWC were found.

2.2. Hydraulic Performance

Only limited information exists in the literature on the hydraulic performance of a non-conventional breakwater-integrated OWC in terms of wave overtopping and transmission. Zanuttigh et al. [31] analysed wave reflections from an OWC device, showing that reflection coefficients are lower than 0.55, which is very low compared to the typical values on vertical caissons that are around 0.9 [32]. Viviano et al. [25] evaluated the influence of the turbine-orifice opening on the wave reflection, to propose an optimisation of the orifice dimension that minimises the reflection coefficient to values lower than 0.6. Recently, Simonetti and Cappietti [33] studied the effectiveness of a the integration of an OWC in a structure as an anti-reflection device to reduce harbour agitation using a numerical wave tank model developed in OpenFOAM®, reaching reflection coefficients of around 15%.

3. Breakwater-Integrated OTD

Another type of WEC suitable to be integrated in both rubble mound and vertical breakwaters is the OTD device (Figure 2). This technology consists of a frontal ramp slope that leads the incident waves to overtop into one or more storage basins (or reservoirs), placed at a level higher than the sea water level. The water stored in the frontal reservoir(s) flows through the turbine(s) combined with a generator that converts the energy from incoming waves into electricity.

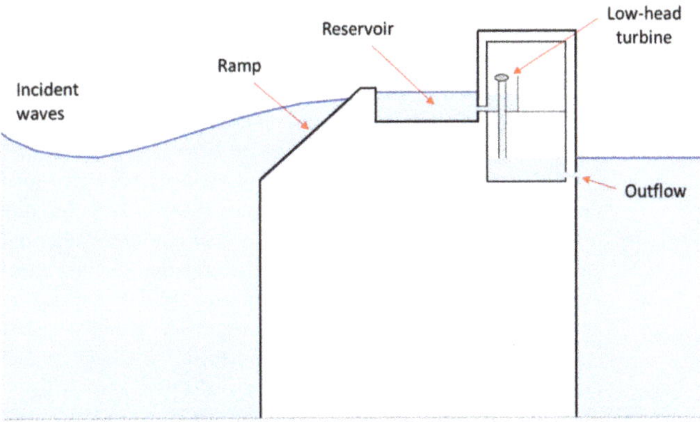

Figure 2. Scheme of the OTD integrated into a vertical caisson breakwater.

The OTD can also be installed on existing breakwaters as demonstrated by [34]. These OTD offers the possibility to store the wave energy, discontinuous in its availability, in the form of relatively stable potential energy. Furthermore, the low head hydraulic turbines used for the energy conversion is a well-established, well-understood and efficient technology.

The biggest disadvantage of this technology is that the energy that can be extracted by the breakwater-integrated OTD device is relatively low from the commercial standpoint. Moreover, the geometry with only one ramp does not appear to be suitable for locations with relevant tidal variations. Seawave Slot-cone Generator (SSG) like systems having two or more ramps have shown to be more efficient regarding energy extraction [35] but, on the other hand, more difficult to handle in prototype-scale breakwaters [36].

A test campaign of an Overtopping BReakwater for Energy Conversion (OBREC) prototype device integrated in an existing rubble mound breakwater is currently ongoing in Italy [6]. This test campaign focus on the wave pressures acting on different parts of the structure, which are being collected during storm events. No information regarding the energy production from the OBREC device is available in literature to date.

3.1. Structural Response/Stability to Waves

The structural response to the wave loading on the breakwater-integrated OTD device has been investigated for many years using physical and numerical model testing. Physical model testing carried out on OTDs with one (OBREC type devices) or more (SSG like devices) ramps have shown that due to their complex geometry, the wave-structure interaction is not comparable to the one observed for traditional breakwaters. Several authors [7,37,38] have shown relevant differences between the forces measured in laboratory on the different parts of the OTD structures and those obtained based on traditional methods used in breakwater design [32]. The performance of a breakwater-integrated OTD has also been investigated by several authors (see e.g., [39–42]) using advanced Computational Fluid Dynamics (CFD) modelling, following which new approaches to evaluate the wave forces and pressure distribution on such innovative breakwaters have been proposed.

3.2. Hydraulic Performance

The hydraulic performance of breakwater-integrated OTDs has been also investigated with the use of physical and numerical model testing. Iuppa et al. [43] and Di Lauro et al. [40] demonstrated that the wave reflection and overtopping at the rear side of a breakwater-integrated OTD with one ramp cannot be evaluated directly considering the formulas proposed in the [43,44] for traditional harbour defence structures. In general, due to the wave dissipation on the ramp and reservoir, the reflection coefficient computed in front of the structure is lower than that computed in front of conventional vertical breakwaters.

Regarding the wave overtopping, results presented by Vicinanza et al. [7] showed that an OTD device integrated into rubble mound breakwaters leads to higher mean overtopping compared to the values measured on traditional breakwaters, due to the presence of a smooth frontal ramp. Contrarily, when the OTD device is integrated into vertical breakwaters, the presence of a setback wall generally reduces the wave overtopping when compared to the typical values observed on conventional caissons. Recently, Iuppa et al. [45] performed additional physical model testing on the OBREC aiming at providing reliable methods for evaluating the energy that can be extracted by said device. In particular, the tests aimed at investigating the probability distribution of the individual overtopping volumes entering the OBREC's reservoir. Cavallaro et al. [46] proposed a numerical model based on the stochastic description of the wave overtopping phenomenon for the optimization of the OBREC device performance.

Specific formulas were presented by Iuppa et al. [43,45] for a rubble mound breakwater-integrated OTD and by Di Lauro et al. [41] for a vertical caisson breakwater-integrated OTD to estimate the reflection coefficient and the mean wave overtopping discharge at the rear side of such non-conventional breakwaters.

4. Hybridisation

As introduced earlier, the hybridisation of different WEC devices and its integration in harbour defence structures appears to be a promising approach to improve the overall efficiency and overcome some of the issues associated with the stand-alone OTD and OWC technologies. The hybridisation has the potential to realistically contribute to the development of WEC technologies in a way that they can compete on the market with more established renewable energy systems, such as wind or solar photovoltaic.

From the research carried out over the last few years on single WEC devices, it is clear that there is an opportunity for the combination of two (or more) WEC technologies into a single system, thus incorporating the advantages of each stand-alone WEC, whilst mitigating their inherent weaknesses. Cappietti et al. [47] presented an experimental study of a concept design of an original HWEC in which the energy is harvested by means of the combination of a frontal OWC and OTD integrated into a vertical caisson, showing that the combination of technologies generally leads to an improved overall efficiency.

The hybridisation is, obviously, not without its challenges. The design of WEC devices integrated in breakwaters is mostly based on several assumptions, which are device specific and cannot be directly extrapolated from one device to another. Assumptions are generally a function of the characteristics of one device and may not necessarily hold true for the hybrid solution. On the other hand, harbour defence structures must accommodate different concepts, which adds complexity in terms of the geometry and integration into the breakwater.

Within the SE@PORTS project, existing WEC devices were assessed on their suitability to be hybridised and integrated in vertical and rubble mound breakwaters (see e.g., [11,12,48]). Several variables/parameters were considered while comparing different preliminary hybrid concepts in multi-criteria analysis: cost-effectiveness; constructability; WEC level of maturity; scalability/modularity; maintenance; reliability; and innovation.

As a result, an innovative HWEC concept combining two well-established principles in harnessing wave energy (OTD + OWC) was designed having as prototypes for the breakwaters and environmental conditions, the ports of Leixões (Portugal) and las Palmas (Spain). These two locations were selected as prototype case studies for several reasons. Firstly, these harbours are protected by breakwaters with a good exposure to incident waves, with maximum significant wave heights that can be higher than 8 m. On the other hand, these harbour facilities have high electricity demands that can be, at least partially, ensured by the harnessed wave energy. Finally, the combination between the local environmental conditions and distinct characteristics of the breakwaters—rubble mound in the case of Leixões and vertical in the case of Las Palmas—make them representative of very many structures protecting EU harbours.

The devised HWEC enables to exploit the strengths of each stand-alone technology and help overcome individual limitations and weaknesses, while presenting a breakthrough and efficient approach to harness wave energy in ports; one that can contribute to increase the capacity for greening the ports.

5. Proof of Concept of the Breakwater-Integrated HWEC

A scheme of the developed HWEC when integrated into a rubble mound breakwater is shown in Figure 3.

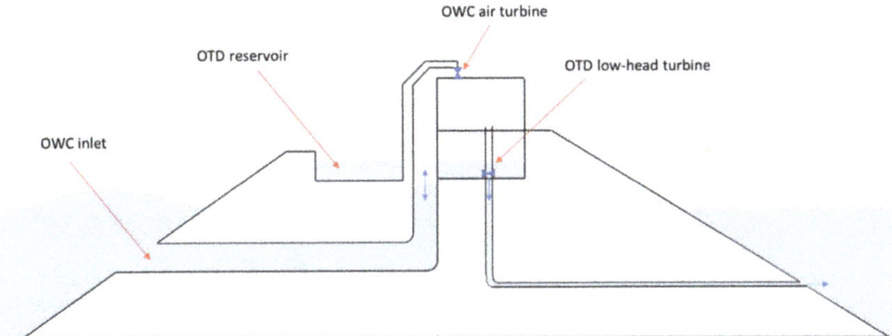

Figure 3. Scheme of the HWEC integrated into a rubble mound breakwater.

In this design, the reservoir of the OTD device is located in front of the OWC chamber. In the OTD device, the water that overtops the reservoir crest produces electricity by passing through low head turbines coupled with generators located in a machinery room placed at the rear side of the breakwater. The water passing through the turbines is then discharged behind the structure and below the still water level via a rear duct. The OWC inlet is located in front of the structure well below the still water level. The oscillation of the water that enters into the OWC chamber drives an air turbine located at the top, as depicted in Figure 3.

From the combined WEC devices, it is possible to harness wave energy over a wider range of metocean conditions using two different yet complementary technologies.

The geometrical optimisation of the innovative HWEC when integrated into a rubble mound or a vertical breakwater and its assessment with respect to its hydraulic and energy performances, were studied through a composite modelling approach [19]. Composite Modelling, defined as *'the integrated and balanced use of physical and numerical models'* [49], is the most advanced approach to study complex problems in marine engineering and design, because it exploits the strengths and overcomes the weaknesses of each approach thereby reducing uncertainties, and therefore is the most suitable to study the breakwater-integrated HWEC.

5.1. Physical Model Testing

Rosa-Santos et al. [12] investigated the influence of the integration of a HWEC module (OWC + OTD with 4 reservoirs) on the structural response/stability and hydraulic performance on the rubble-mound cross-section proposed for the extension of the north breakwater of the Port of Leixões in Portugal. Experimental tests of the hybrid device were performed on a geometrical scale of 1:50 in the wave basin of the Faculty of Engineering of the University of Porto (FEUP) in Portugal, using both regular and irregular waves. This wave basin is equipped with a multi-element wave generation system from HR Wallingford. Wave reflections were minimised by a dynamic reflection absorption system integrated in the wavemaker [50]. The duration of the tests was long enough to contain around 100 waves for regular wave tests and around 1000 waves for tests with irregular sea state [12].

At an early stage, experimental testing under regular wave conditions was carried out to investigate three geometries (A, B and C) with different OWC inlet configurations and identify the most efficient among those. The study focused on the evaluation of the wave reflection from the structure, scour development and toe stability, and average wave overtopping discharge over the crest of the innovative structure. Preliminary results by Rosa-Santos et al. [12] highlighted the hydraulic performance of the HWEC integrated into the rubble-mound breakwater. The authors showed that wave reflection was not significantly affected by the integration of the HWEC module, for which values of the reflection coefficient around 0.35 were found. Furthermore, the presence of the HWEC leads to a reduction of the transmitted wave by overtopping at the rear side of the structure compared to the conventional cross-section, thus improving the functional performance of the breakwater. The experimental tests also indicate that the integration of the HWEC in the rubble-mound breakwater does not significantly affect the stability of the toe berm and does not contribute to the further development of scour in front of the structure. Overall, these preliminary results indicate that the integration of the HWEC in rubble-mound breakwaters can be a promising solution, as it does not seem to compromise its stability nor functionality as a harbour defence structure.

The structural response/stability and hydraulic performance of the vertical caisson breakwater-integrated HWEC was tested in model scale in the wave flume of the University of Cantabria [48] with an optimised geometry suitable to be integrated in the Nelson Mandela breakwater at Port of Las Palmas in Spain. Different configurations of this innovative structure were tested with detailed evaluations of the operational performance of the combined devices under mild conditions, as well as calculations of the wave forces on the structure for the stability and functionality assessment of the breakwater under extreme conditions. Preliminary results by Lara et al. [47] indicate that the combination of a one reservoir OTD with a OWC reduces the reflection coefficient to a value up to 0.3 for the cases where the wave period coincides with the resonant period of the OWC due to the large amount of water flowing inside the OWC chamber. This result represents a remarkable improvement of the innovative breakwater compared to vertical caisson breakwaters, as the typical values of the reflection coefficient for these are around 0.9 [32]. Therefore, the study by Lara et al. [48] suggests that the HWEC integration can improve the hydraulic performance of the breakwater, reducing thereby e.g., problems for navigation caused by reflected waves.

The performance of the HWEC in terms of energy production have been presented by Cabral et al. [50,51] by analysing the data of the experimental tests of the three different alternative HWEC geometries (A, B and C) when integrated in a rubble mound breakwater constructed to a 1:50 geometrical scale. As mentioned earlier, the only difference between these geometries was the design of the OWC chamber inlet. The authors analysed the hydraulic efficiency, i.e., the ratio between the average power absorbed by the device and the average power of the incident waves in front of the structure, of the HWEC device and the stand-alone OTD and OWC devices installed into the breakwater for each of the three alternatives. The analysis showed that the combination of the OTD with the OWC leads to higher hydraulic efficiencies than those typically reported in the literature for the independent components, for a broader range of hydrodynamic conditions, reaching a maximum overall hydraulic efficiency of around 40% for geometries B and C [50,51]. Due to the combination of different WEC systems, the overall performance is improved since the two selected technologies complement each other well, thus extending the range of wave conditions where the efficiency is high. In detail, the multiple reservoir OTD was shown to be efficient for the lower wave periods, while the opposite occurs for the OWC system for all three alternatives (A, B and C). Consequently, the range of hydrodynamic conditions—i.e., waves and tidal levels—for which the device's power production is low was reduced. This is an important advantage of the HWEC device because it allows a more constant power production with a significant reduction of the power production peaks, which is usually difficult to obtain in the renewable energy sector.

5.2. Numerical Modelling

In parallel with the physical model testing conducted in the wave basin at FEUP (see e.g., [12,50,51], two-dimensional numerical modelling using ANSYS Fluent CFD model [52] has been performed as well for identifying the most efficient HWEC geometry and then using it for performing additional tests under various wave and water level conditions.

The geometry of the numerical model consisted of a wave flume, a breakwater, and the integrated HWEC device selected according to the physical model tests conducted in Portugal. Firstly, the model was validated with the preliminary data of the experimental tests assuming a specific geometry for the HWEC (Geometry B). Then, Geometry B was compared numerically with the alternative Geometry C from which the most efficient geometry was identified, thereby confirming the outcome of the experimental study. Finally, the most efficient geometry was adopted to examine more sea state conditions that were not tested in the laboratory. The numerical analysis focused on the evaluation of: (a) the free surface elevation in various locations along the wave flume and at the breakwater toe; (b) the water level oscillations and the amplification coefficient inside the OWC chamber; and (c) the wave overtopping discharges in the OTD multiple reservoirs.

Said parameters were perceived as being the most relevant to investigate the performance of the innovative HWEC device during this proof of concept stage. In the future, additional simulations will be performed involving three-dimensional models of the breakwater-integrated HWEC for further evaluation of its structural response/stability and hydraulic performance and the performance of the HWEC in harvesting wave energy, as well as to determine specific and more detailed parameters necessary in the design of the turbines.

5.2.1. Model Set-Up

As already mentioned, a two-dimensional numerical model was implemented using the commercial software ANSYS Fluent. For the simulations, the two-dimensional Reynolds-Averaged Navier-Stokes (RANS) equations for incompressible flow were employed, while for turbulence closure the shear-stress transport (SST) k-ω turbulence model was chosen. For the free surface elevation tracking the Volume of Fluid (VOF) method was used. The VOF formulation is based on the assumption that two or more fluids are immiscible. In open-channel flow (as is the case being tested), the governing equations across the water–air interface are expressed as a single fluid whose physical properties are defined by the volume fraction weighted average of the corresponding physical properties of air and water, i.e., for density $\rho = \alpha\, \rho_{water} + (1 - \alpha)\, \rho_{air}$, where α is the water volume fraction. In a finite volume method (ANSYS Fluent), α is defined in each computational cell as $\alpha = \delta V_{water} / \delta V_{cell}$, where δV_{cell} is the volume of the computational cell and δV_{water} is the volume of water in the cell. Therefore, $\alpha = 1$ if the cell is full of water, $\alpha = 0$ if the cell is full of air or $0 < \alpha < 1$ if the cell contains the water–air interface, respectively.

The model was set-up to replicate exactly the mid-transverse cross section of the wave flume geometry and HWEC designed and tested in the laboratory (see e.g., [50,51]). As already mentioned, one geometry of the HWEC (Geometry B from Cabral et al. [50]) was used for validation purposes.

The numerical domain was 15.80 m long and 1.82 m high (Figure 4). The water level in the numerical domain was kept constant for all the validation tests, with a water depth of 0.488 m at the left boundary of the domain (offshore). This water depth represents the value of the Mean Water Level (MWL) at the port of Leixões Port, constructed to a 1:50 geometrical scale.

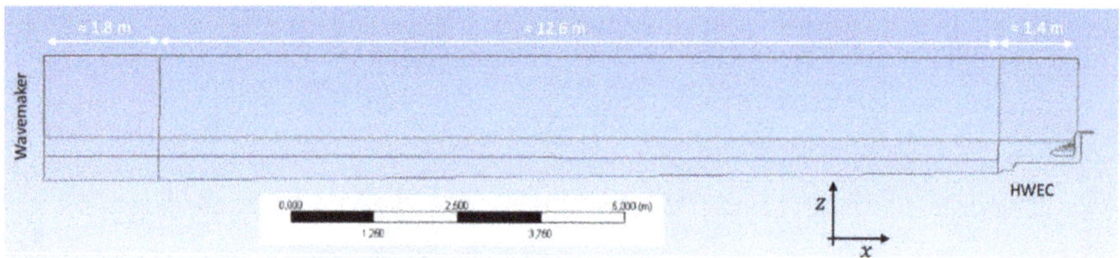

Figure 4. Numerical model domain.

The numerical domain was divided into several sections having different mesh resolutions: three different zones in the longitudinal direction and two to three zones in the vertical direction as indicated in Figure 4. The three different regions in the longitudinal direction were: the region close to the wavemaker (around 1.8 m long), the central region covering the main channel region (around 12.6 m in length) and the one around the HWEC model (around 1.4 m long). The vertical levels corresponding to the water region on the bottom (around 0.46 m high), the region near the free surface (around 0.16 m high) and the air region on the top (around 1.2 m high). The final height of the air region was selected after following a trial and error approach until the top boundary stops affecting the results in the water region.

The computational grid had a non-uniform size. From a sensitivity analysis on the mesh resolution, it was decided that the mesh size in the region near the free surface and close to the HWEC device was where the highest resolution was needed. The mesh size in this region was 0.003 m. For the top region (air region) a lower resolution was set with mesh sizes varying from 0.003 m to 0.036 m. The other regions of the numerical wave flume were set with mesh sizes varying from 0.003 m to 0.018 m. The refined mesh generated around the area of the HWEC is shown Figure 5. The entire mesh contained around 750,000 elements and its overall quality was controlled by keeping the maximum skewness below 0.8.

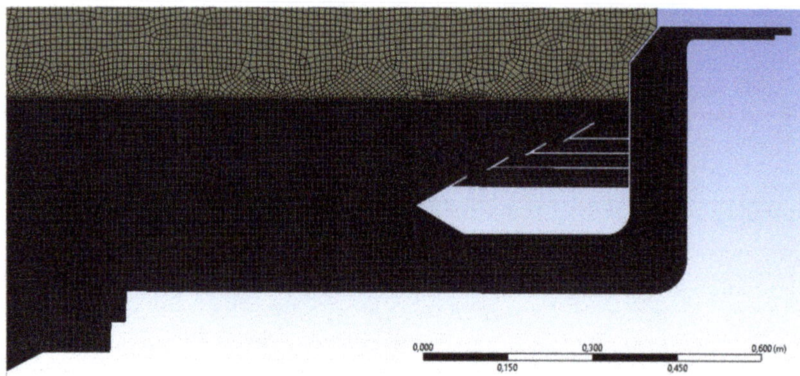

Figure 5. Mesh details near the HWEC device.

The HWEC device consists of an OTD device with four reservoirs (Figure 6). The crest heights of the reservoirs were set at 0.015 m, 0.040 m, 0.065 m and 0.100 m (model scale) from the bottom reservoir (R1) to the top one (R4), respectively. An angle of 30° relative to the horizontal was chosen for the ramps of the OTD device, aiming at not only ensuring the occurrence of slightly breaking surging waves, which produce low energy dissipation [10], but also approximately preserving the original slope of the breakwater extension of the Port of Leixões (Portugal), equal to circa 27°. Note here that the geometry of the ramps, reservoirs, and OWC chambers tested with the CFD is the same as the Geometry B tested in laboratory by Cabral et al. [50].

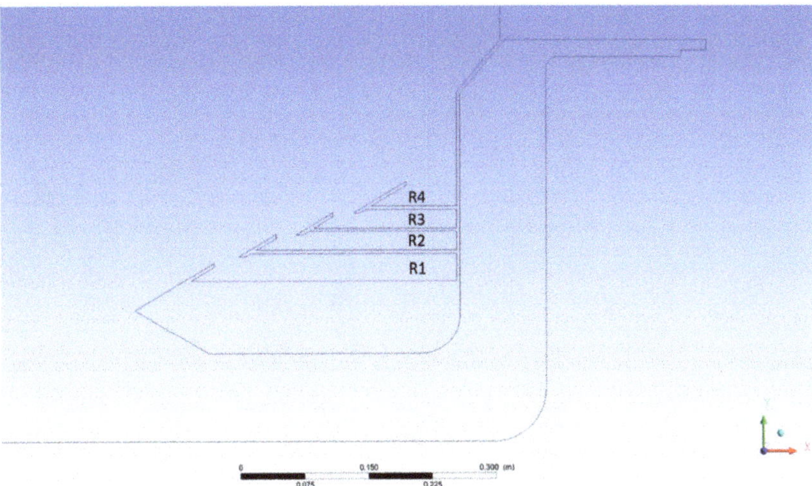

Figure 6. OTD Reservoirs.

In the numerical model, the following discretization schemes were utilised: Pressure Staggering Option (PRESTO) for pressure; Second Order Upwind for momentum; Compressive for the VOF implementation; First Order Upwind for turbulent kinetic energy k; and specific dissipation rate ω and Least Squares Cell Based for the Gradients. The Semi-Implicit Method for Pressure Linked Equations (SIMPLE) scheme was utilised for the velocity–pressure coupling. For the transient discretization a Second Order Implicit scheme was used. The time step in all simulations, which was constant and equal to 6.25×10^{-3} s, was selected so that the CFL (Courant-Friedrichs-Lewy) criterion was always satisfied.

In order for the numerical model to reproduce the wave field measured in the laboratory and the HWEC-wave interaction, several boundary conditions were considered:

- In the laboratory a piston-type wavemaker with an active absorption system was used at the offshore boundary of the wave tank for the generation of various regular and irregular wave conditions. Likewise, in the numerical model a moving wall at the left offshore boundary of the computational domain was implemented to replicate the exact wavemaker motion measured during the experimental tests. Thus, regular and irregular wave time series were generated in FLUENT using the corresponding wave paddle motion measured in the laboratory. This movement was transferred to the model using the measured wave paddle position over time, X(t), and the associated velocity, V(t). This procedure allows replicating wave-by-wave conditions tested in the laboratory (see A in Figure 7).
- Symmetry for the top boundary of the domain (see B in Figure 7) since it has been confirmed that the length of the air region above the water is big enough so not to affect the results in the water region.
- A pressure outlet for the boundary in the air phase just above the HWEC device (see C in Figure 7).
- A fan boundary condition for the OWC (see A in Figure 8).
- A pressure outlet for the sinks of the OTD devices (see B in Figure 8).
- No slip conditions at the bottom, breakwater toe and all other parts of the HWEC device.

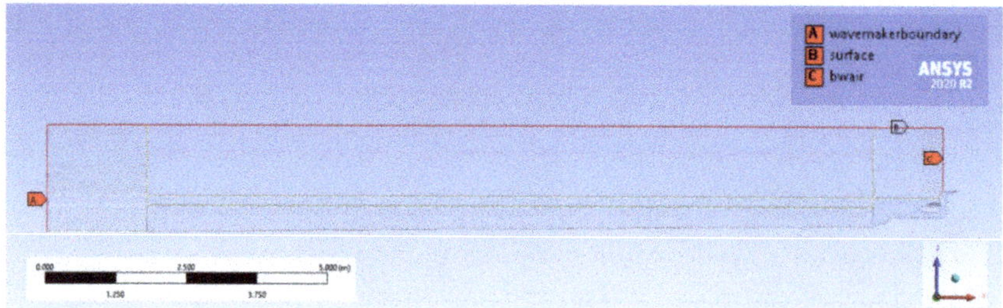

Figure 7. Wavemaker, top surface and air boundary of the computational domain.

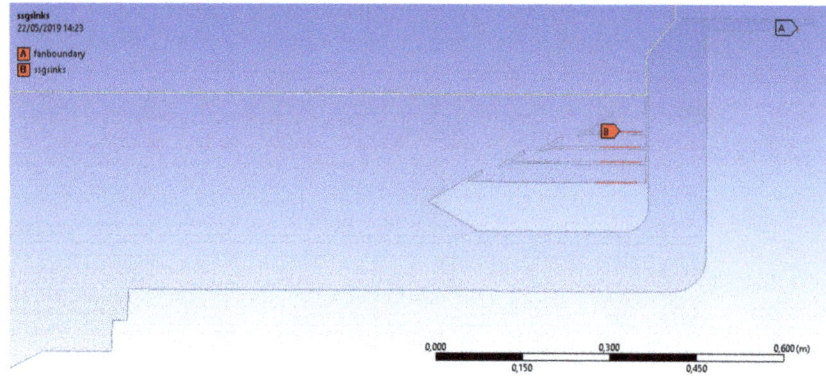

Figure 8. Sinks and Fan Boundaries.

5.2.2. Model Validation

Numerical data were validated against data measured in FEUP's laboratory for three different regular and irregular wave conditions for Geometry B [50]. The waves selected to be validated (in model scale) correspond to the most representative wave conditions in the Port of Leixões in Portugal. The variables compared with the experimental data were the free surface elevation at different locations along the wave flume, the variation of the water level inside the OWC chamber and the mean overtopping discharges over the OTD reservoirs' crest. The experimental setup in the wave basin at FEUP is presented in Figure 9. It is reminded, that the wave tank had a wavemaker at the left boundary and an active absorption system for dealing with reflection issues, as mentioned above (Section 5.2.1 Model Set-up).

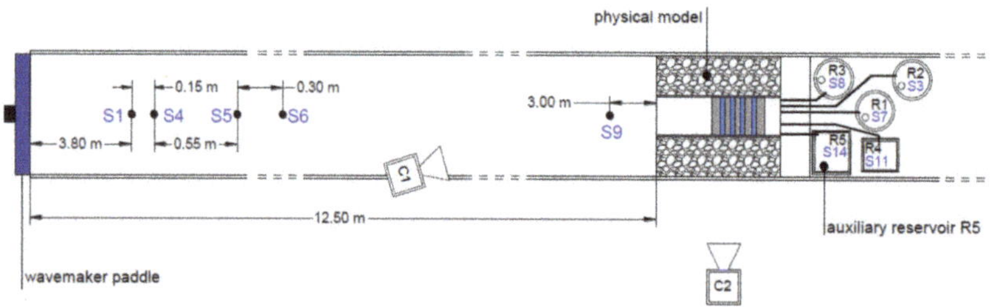

Figure 9. Physical model setup in the channel built inside the wave basin at FEUP [12].

Whereas Figure 10 shows the comparison between the measured and calculated values of the water free surface elevation at two locations (S4 and S5) inside the wave flume (Figure 9), Figure 11 shows the water level comparison inside the OWC chamber. These results refer to a test with regular waves (Val_Test_01) of a wave height, *H*, equal to 0.02 m and a wave period, *T*, equal to 1.7 s. The offshore water depth was equal to *h* = 0.488 m. Thus, the time series presented in Figures 10 and 11 correspond to 15 wave periods, which is an adequate simulation time taking into account the big computational cost of the simulations and the limited variations of the free surface elevation. The small oscillations of the free surface elevation that appear in Figure 10 are attributed to the expected reflection from the HWEC device itself. Nevertheless, it can be concluded that the numerical model is able to represent satisfactorily well the water free surface elevation at the two locations in the wave flume. Regarding the comparison presented in Figure 11 it is clear that the model captures almost precisely the water level oscillations measured inside the OWC chamber.

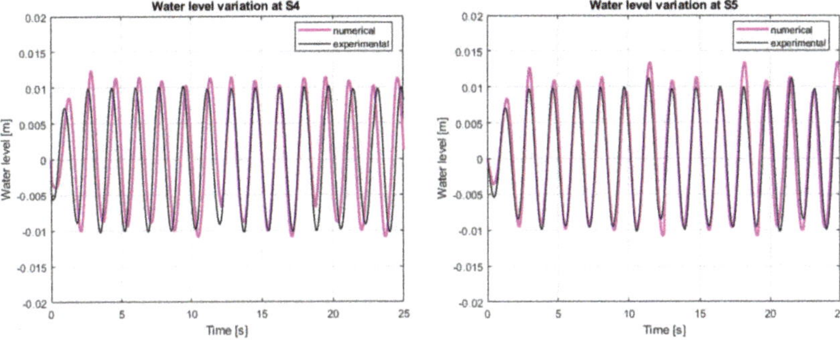

Figure 10. Comparison between measured and calculated water free surface elevation at two locations in the wave flume: wave gauge S4 (**left** panel) and S5 (**right** panel) for Val_Test_01.

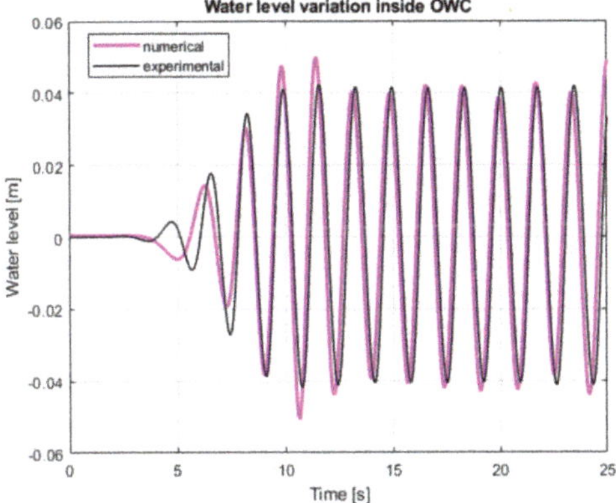

Figure 11. Comparison between measured and calculated water level inside the OWC chamber for Val_Test_01.

The second phase of the validation process included the comparison of the wave overtopping. It is worth noting that in the numerical model, the discharges entering each reservoir were estimated by considering the total amount of wave discharge overtopping the crest of each reservoir. Contrarily, in the physical model testing, the overtopping in the reservoirs was indirectly assessed by measuring the volume of water collected in auxiliary reservoirs located behind the model and connected with each reservoir through rigid pipes (Figure 9). Therefore, the overtopping discharges in the HWEC reservoirs were estimated using the measurements of the water level gauges installed in each of these auxiliary reservoirs. From the variation of the water level inside each auxiliary reservoir, the volume variation was computed from the cross-sectional area of the auxiliary reservoir.

The wave overtopping in the different reservoirs, represented in terms of the cumulative mass flow (kg/m) derived from the numerical model, was compared with the volume variation in the auxiliary reservoirs derived from the experiments and translated into mass flow (kg/m), as depicted in Figure 12. Results refer to a regular wave test (Val_Test_02) with $H = 0.08$ m and $T = 1.7$ s. The offshore water depth was equal to $h = 0.488$ m. Reservoirs are numbered from R1 to R4, where R1 is the lowest and R4 the highest one in the OTD system. Results shown in Figure 12 indicate that the model predicts accurately the mass flow in reservoirs R2, R3 and R4, while it slightly overestimates the mass flow in the lowest reservoir R1. This slight overestimation was expected since the first reservoir was partially saturated during the physical model test due to the low crest freeboard of the first ramp.

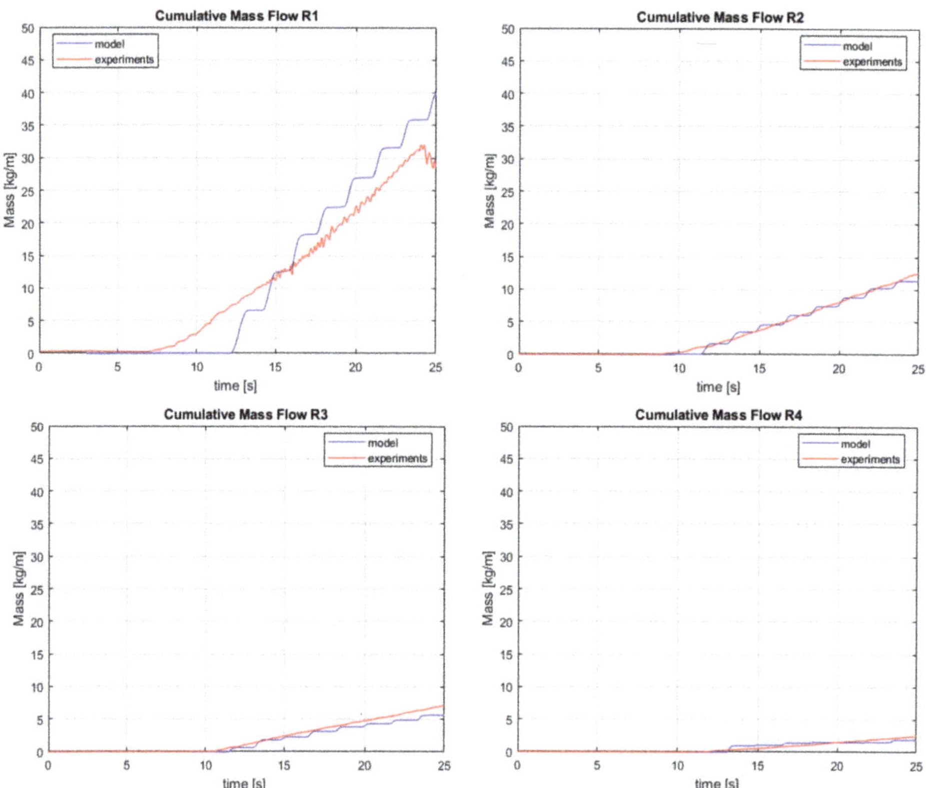

Figure 12. Comparison between measured and calculated cumulative mass flow (kg/m) in the four reservoirs of the HWEC for Val_Test_02.

For this validation test a snapshot of the free surface elevation during a time instant after 14 wave periods is presented in Figure 13, while the same snapshot but with a focus near the HWEC device is presented in Figure 14, respectively. It can be verified that the overtopping is less pronounced in the reservoirs R2 and R3 comparing with the lowermost R1 reservoir, while the overtopping in the uppermost R4 reservoir is very small.

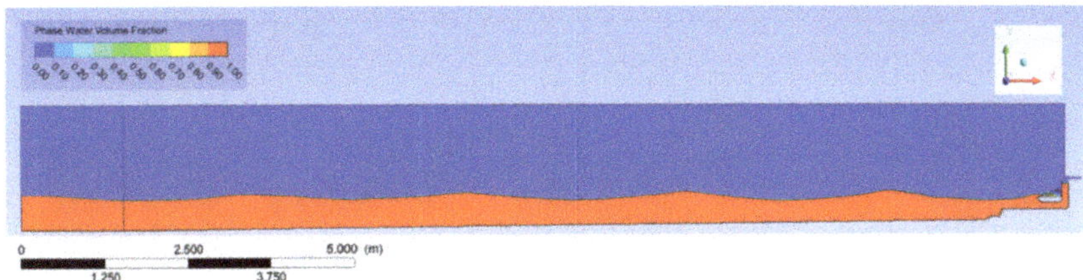

Figure 13. Snapshot of free surface elevation at a time instant after 14 wave periods in the entire domain.

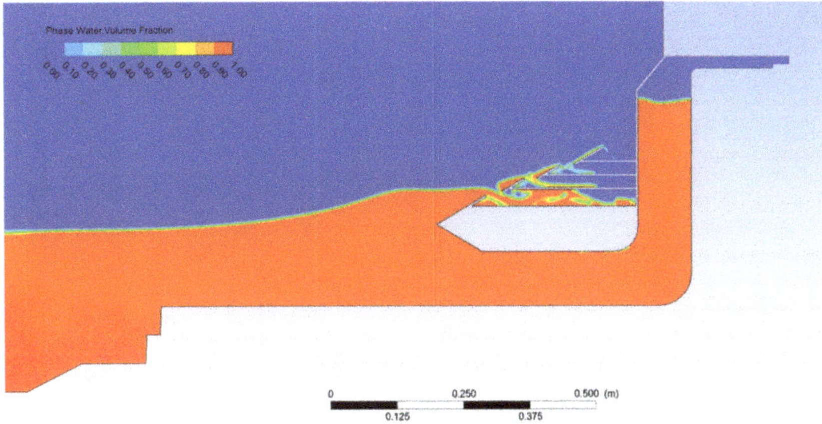

Figure 14. Snapshot of free surface elevation at a time instant after 14 wave periods close to the HWEC device.

The third test used for the validation of the model corresponds to an irregular wave test (Val_Test_03) with a significant wave height, H_s, equal to 0.034 m and a peak wave period, T_p, equal to 1.27 s. The offshore water depth was equal to $h = 0.488$ m. The simulation time was equal to 30 wave periods, two times bigger than in regular wave cases since we wanted to ensure that possible variations occurring are captured. The comparison between the measured and calculated values of the free surface elevation at two locations (S1 and S9) in the wave flume is presented in Figure 15. These results show that the numerical model captures satisfactorily well the water free surface elevation also for the case of irregular waves, although some discrepancies can be observed. Figure 16 displays the comparison between measured and simulated oscillation of the water surface inside the OWC chamber for Val_Test_03. It is noted that the model is able to capture adequately the water level oscillations inside the chamber for the first 45 s, which correspond to around 26 incident waves. After around 45 s the comparison is not as good. This is attributed to the fact that the model is two-dimensional and therefore not able to capture three-dimensional phenomena occurring in the physical model testing. In the laboratory experiments it was observed that after some time the three-dimensional effects were not insignificant inside the OWC. More specifically, inside the OWC chamber waves in the transverse direction were created. These waves are responsible for the unsteady pattern of the black wave signal observed after around 45 s in Figure 16, which cannot be captured by the numerical model. Regarding the wave flume in front of the device the three-dimensional effects during the experiments were not that pronounced after the 45 s, thus the comparison with the numerical results presented in Figure 15 is much better.

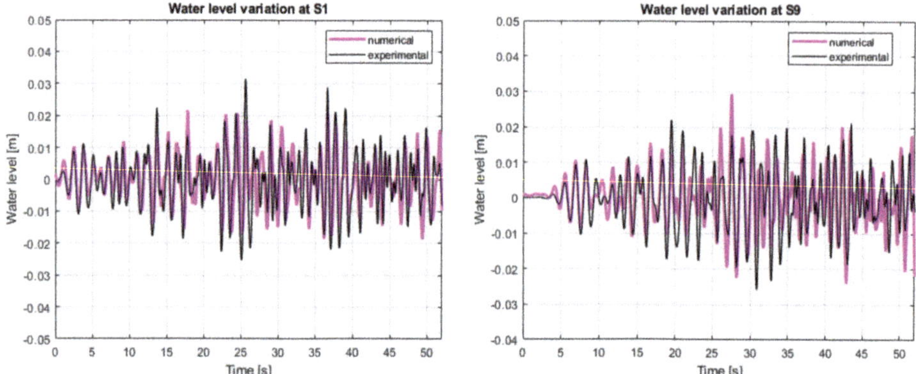

Figure 15. Comparison between measured and calculated free surface elevation at two locations in the wave flume: wave gauges S1 (**left** panel) and S9 (**right** panel) for Val_Test_03.

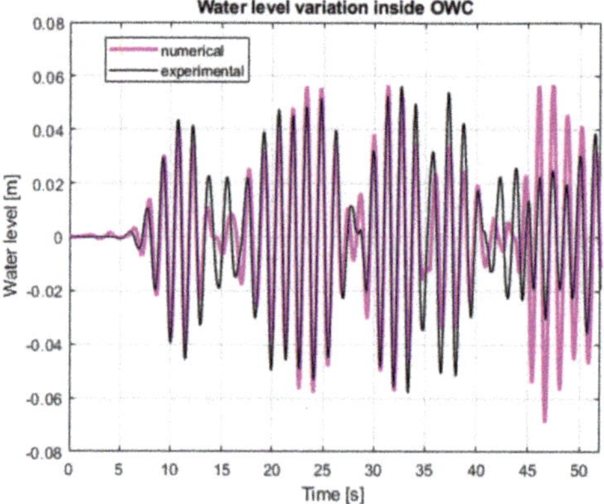

Figure 16. Comparison between measured and calculated water level inside the OWC chamber for Val_Test_03.

Table 1 indicates the Root-Mean-Square Error (RMSE) of the free surface elevation between measured and calculated data for the five wave gauges locations along the wave flume (as shown in Figure 9) and inside the OWC chamber for the three tests described before. The RMSE was calculated according to the equation presented herein below. Although discrepancies can be noted due to the complex interaction between the waves and the non-conventional geometry of the HWEC, results of this validation indicate that the numerical model is capable of sufficiently reproducing the wave field inside the wave flume as well as the oscillation inside the OWC chamber for all cases tested.

$$\mathrm{RMSE}(\%) = sqrt\left(\frac{\frac{1}{N}\sum_{i=1}^{N}\left(\left(H_{numer} - H_{exper}\right)^2\right)}{\frac{1}{N}\sum_{i=1}^{N}\left(H_{exper}^2\right)}\right) * 100$$

Table 1. RMSE (%) of the water level oscillation calculated with the numerical model when compared to the measured data.

Test	Wave Gauge S1	Wave Gauge S4	Wave Gauge S5	Wave Gauge S6	Wave Gauge S9	Wave Gauge in OWC Chamber
Val_Test_01	4.9	4.8	2.1	5.2	5.3	8.1
Val_Test_02	5.6	5.2	5.4	5.8	6.1	10.3
Val_Test_03	5.9	5.4	5.7	6.8	7.3	12.7

5.2.3. HWEC Geometry Optimisation

After the model validation, two alternative geometries (B and C) designed to be tested in the laboratory were also compared through numerical model simulations. From the initial testing, these two geometries seemed to be the most efficient among the three alternatives. The scope of the additional numerical simulations was to identify the most effective geometry in terms of wave overtopping in the multiple reservoirs and water oscillation inside the OWC chamber, along with providing an additional tool for geometry optimisation that is complementary to the physical model testing. In this section, the numerical results for the HWEC geometry optimisation are discussed. As noticeable in Figure 17, the two geometries are the same except for the shape of the OWC chamber inlet.

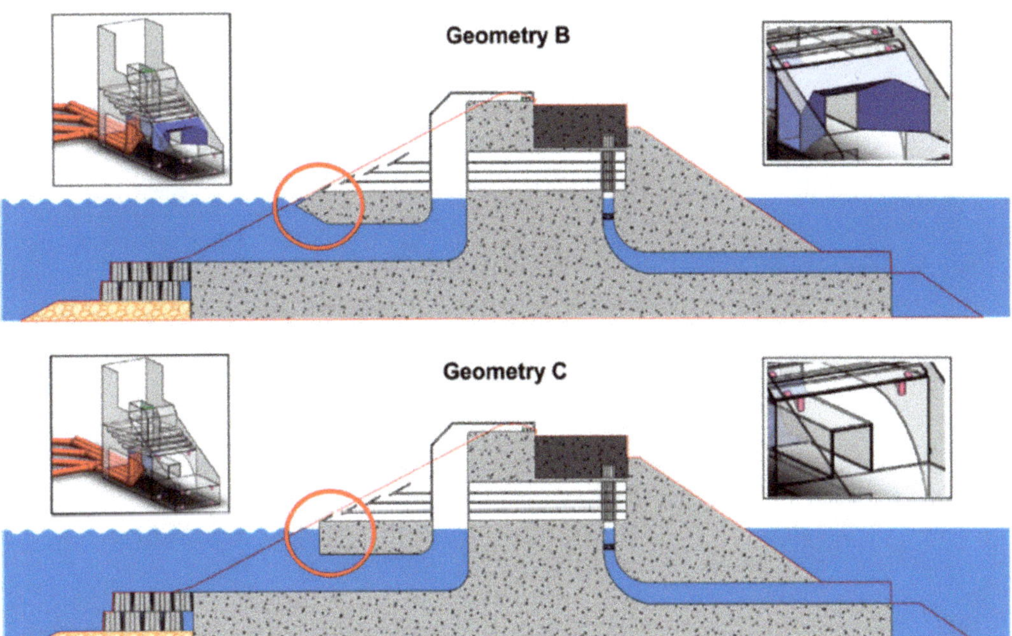

Figure 17. Sketch of the two different geometries of the HWEC system analysed for the geometrical [50].

The characteristics of the numerical model domain, boundary conditions and numerical settings were the same as the ones adopted for the model validation and described in previous sections. Details of the mesh for the two geometries are presented in Figure 18. Note here that the same mesh discretisation near the HWEC and all over the domain was set for the two geometries.

The two configurations were tested considering the same incident irregular wave conditions, with $H_s = 0.034$ m, $T_p = 1.27$ s and offshore water depth, $h = 0.488$ m. These wave characteristics are the ones corresponding to the "Val_Test_03".

In Figure 19 the free surface elevation at three locations (S1, S5 and S9) in the numerical wave flume is displayed for the two configurations, while Figure 20 shows the comparison of the water level oscillation inside the OWC.

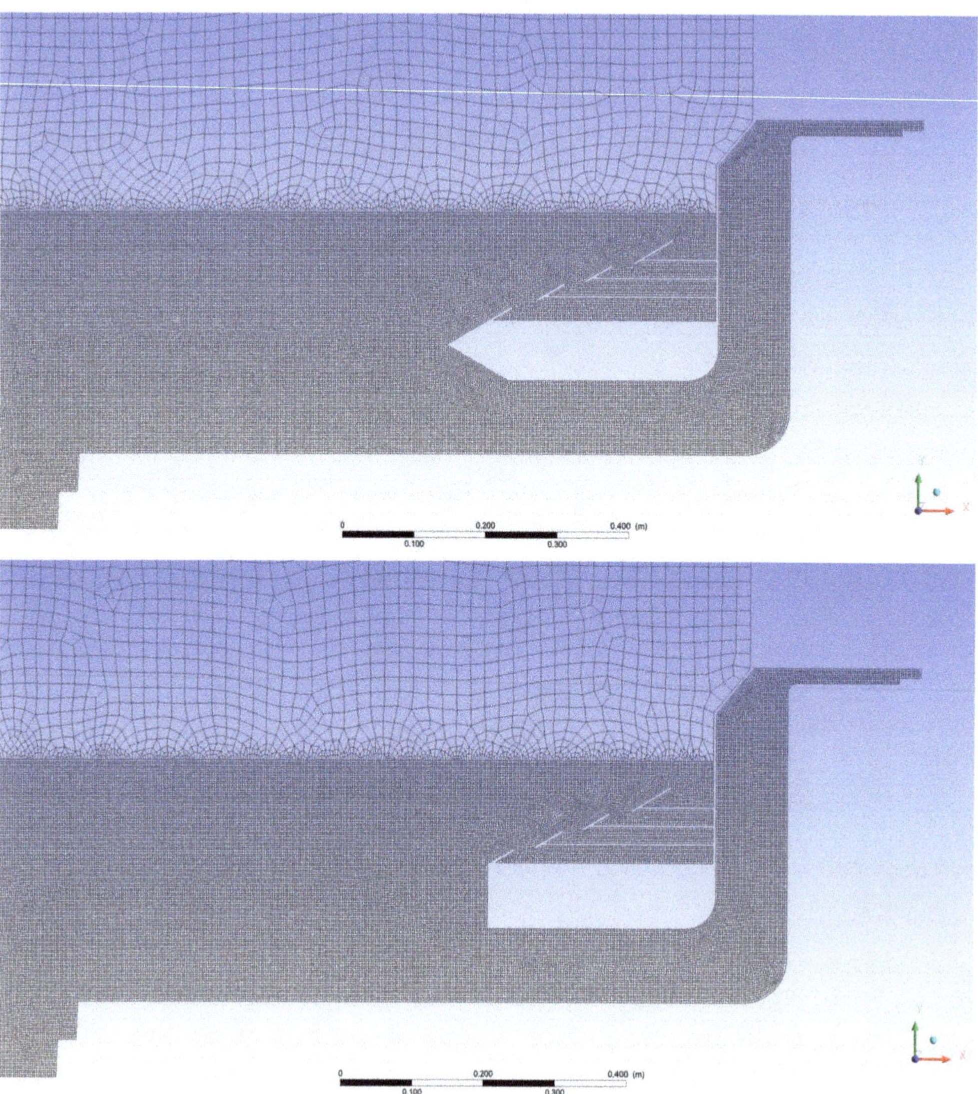

Figure 18. Mesh details near the HWEC device for Geometry B (**upper** panel) and C (**lower** panel).

Regarding the wave field in the wave flume, results indicates that the two different configurations have no significant effect on it and only some small differences of the free surface elevation can be observed. This behaviour was later verified in the physical model testing. In the experiments the difference in the reflection coefficient between the two geometries was small. Results on the water oscillation inside the OWC between the two geometries are also very similar, as shown in Figure 20, though slightly better for Geometry B.

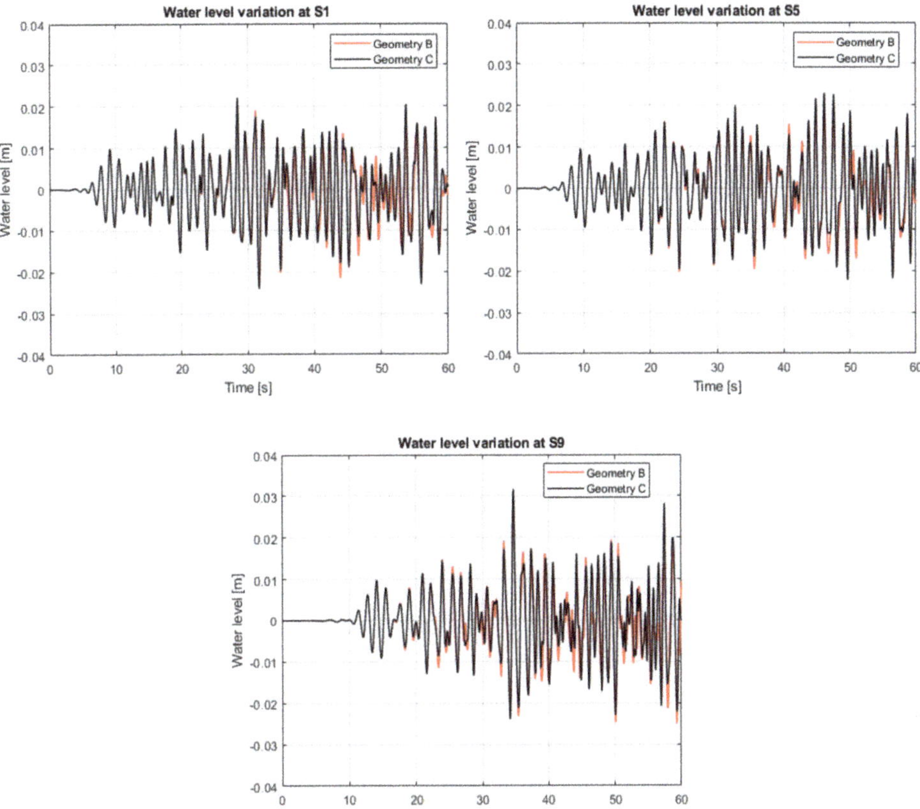

Figure 19. Comparison of the free surface elevation between the two geometries at locations: S1 (**top left** panel), S5 (**top right** panel) and S9 (**lower** panel) in the numerical flume.

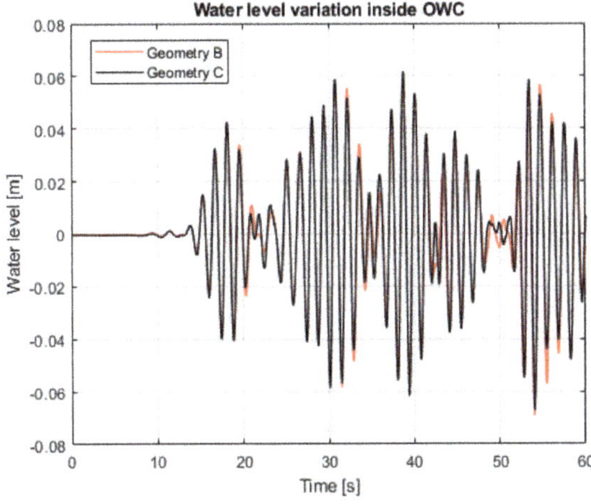

Figure 20. Comparison of the water level oscillation inside the OWC chamber for the two geometries.

Regarding the overtopping discharges into the different OTD reservoirs, translated into mass flow (kg/m), bigger differences between the two geometries are observed (Figure 21). In the two lowest reservoirs (R1 and R2), in which the discharge is significant, larger wave overtopping can be noted for Geometry B. This configuration is much more efficient than Geometry C, with wave overtopping larger in Geometry B than the one calculated for C. This is an expected response according to Cabral et al. [50] since the special ramp near the OWC entrance channel in Geometry B was designed for leading more overtopping into the reservoirs. For the upper two reservoirs (R3 and R4) the amount of overtopping is very small, so the difference between the two geometries is not well captured. According to these results, the performance of the Geometry B HWEC was better than Geometry C HWEC, mostly due to the larger computed wave overtopping into the OTD reservoirs. This preliminary finding obtained by the CFD modelling was also confirmed later by Cabral et al. [50] when testing the different geometries in laboratory. More specifically, for the same wave period tested, the hydraulic efficiency of Geometry B was greater compared to the one of Geometry C.

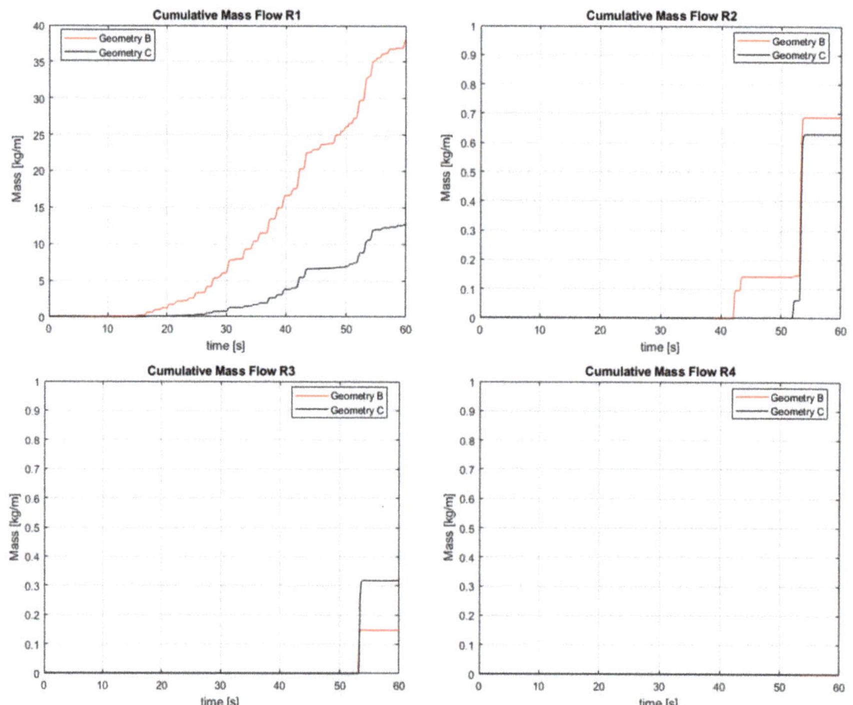

Figure 21. Comparison of the cumulative mass flow (kg/m) in the four reservoirs between the two geometries.

5.2.4. Hydraulic Performance

Once the results of the numerical model were validated against the measured data and the most effective geometry was identified, additional numerical simulations with the most effective geometry (Geometry B) were performed in order to extend the physical model test programme with wave characteristics and water level conditions not tested in laboratory. The additional simulations were performed assuming irregular waves on the Froude scale model of the prototype breakwater at the Port of Leixões constructed to a 1:50 geometrical scale. Table 2 summarises the significant wave height, H_s, peak period, T_p, and offshore water depth, h, in the additional numerical tests.

Table 2. Main characteristics of the additional numerical model tests.

	Model Scale			Prototype Scale		
Test	h (m)	H_s (m)	T_p (s)	h (m)	H_s (m)	T_p (s)
Reference test	0.488	0.034	1.27	24.40	1.70	9
Test_01	0.461	0.034	1.27	23.05	1.70	9
Test_02	0.475	0.034	1.27	23.75	1.70	9
Test_03	0.504	0.034	1.27	25.20	1.70	9
Test_04	0.520	0.034	1.27	26.00	1.70	9
Test_05	0.488	0.055	1.27	24.40	2.75	9
Test_06	0.488	0.075	1.27	24.40	3.75	9

It is well known from the literature that tide can have a great (negative) impact on the energy production, in particular on the OTD device. Therefore, in order to evaluate the effect of the still water level on the performance of the HWEC, four additional different offshore water depths were tested in the numerical wave flume (Test_01 to Test_04 in Table 2). In addition, in order to investigate the influence of the significant wave height H_s on the wave overtopping discharges and water oscillation inside the OWC chamber, two additional cases were examined (Test_05 and Test_06 in Table 2). In these two cases, the still water level was set equal to the level tested in laboratory (h = 0.488 m), representing the MWL in prototype scale in front of the Port of Leixões. Note that all the additional numerical tests were carried out considering a constant peak wave period T_p = 1.27 s.

The effect of the offshore water depth was initially investigated considering the results of the overtopping discharges into the four reservoirs of the HWEC device. As expected, as the still water level increases, the cumulative overtopping discharge, translated in mass flow (kg/m), inside the reservoirs increases, as displayed in Figure 22. This is noticeable in the lowermost reservoir (R1), as the waves overtop the crest and reach the corresponding sink in all cases. In the second and third reservoirs (R2 and R3) overtopping occurs only in the cases where the still water level is above the mean water level (Test_03 and Test_04) and, especially in the third reservoir, the discharge is not significant. Almost no overtopping is observed in the uppermost sink.

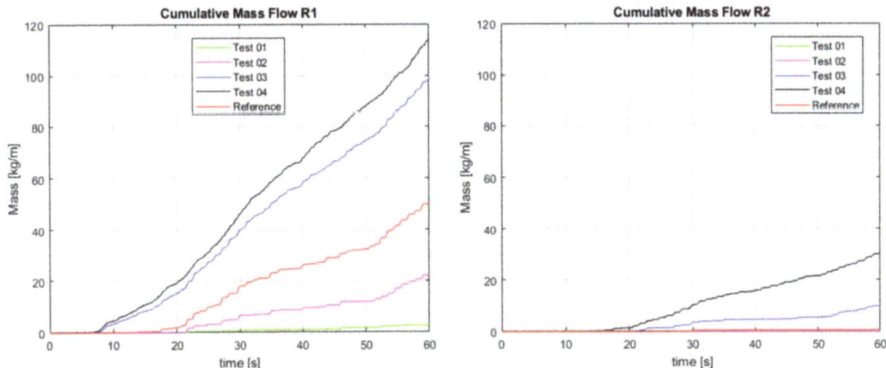

Figure 22. *Cont.*

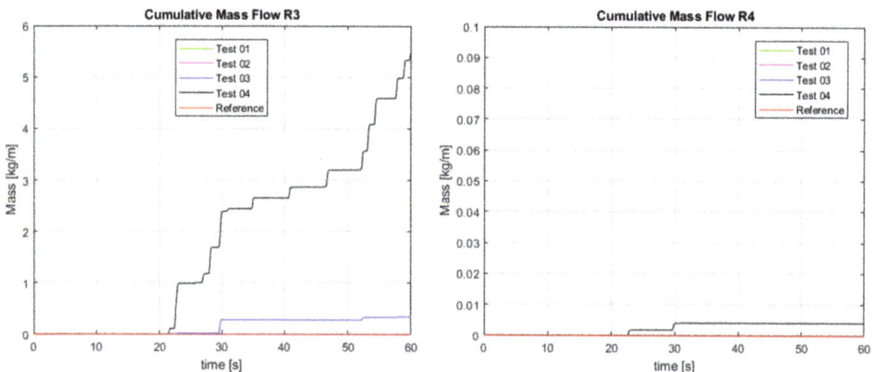

Figure 22. Comparison of the cumulative mass flow (kg/m) in the four reservoirs between the Reference test and Test_01 and Test_04. Note that the y axis is on a different scale at the four reservoirs for comparison purposes.

The effect of H_s on the wave overtopping discharges in the reservoirs was also investigated with Test_05 and Test_06. This is clearly shown in Figure 23 for the lowest reservoir (R1), which is continuously overtopped for almost every incident wave, with larger water discharges calculated. In the second and the third reservoirs (R2 and R3) the trend is similar, but the discharges are smaller. Finally, in the uppermost reservoir (R4) overtopping occurs only in the case of the larger significant wave height (Test_06).

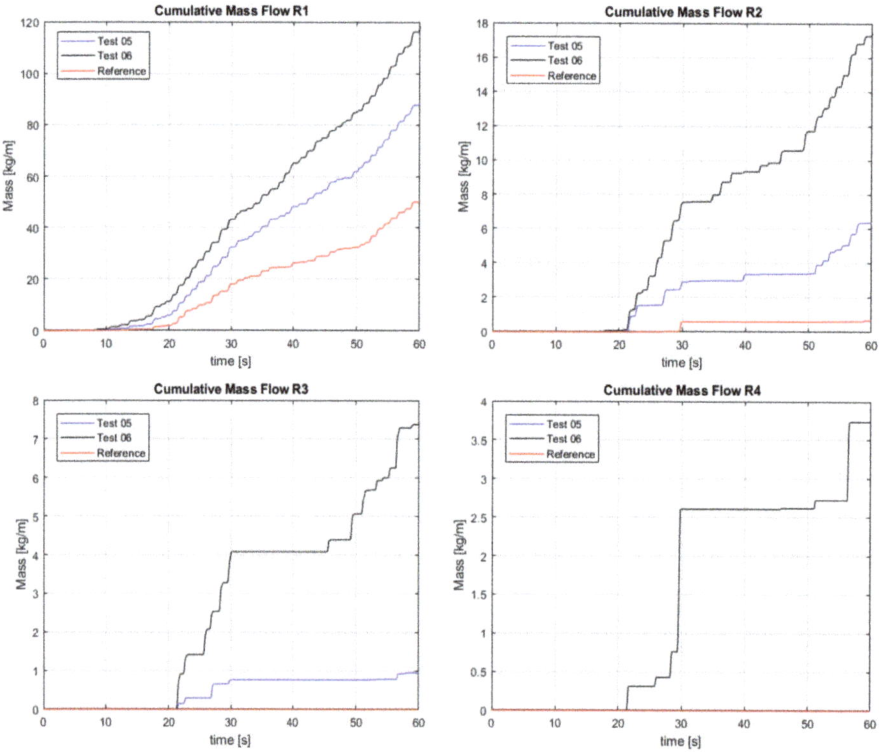

Figure 23. Comparison of the cumulative mass flow (kg/m) in the four reservoirs between the Reference test and Tests_05 and Tests_06. Note that the y axis is on a different scale at the four reservoirs for comparison purposes.

Regarding the effect of the offshore water depth on the water level oscillation inside the OWC, results show that an increase in the still water level leads to a decrease in the water oscillations inside the OWC (Figure 24—left panel). As can be noted, highest oscillations occur for Test_01 which is characterised by the smallest still water level. The same trend is observed in Figure 24—right panel which presents the effect of the still water level on the amplification coefficient. The amplification coefficient has been derived by dividing the free surface elevation with the significant wave height, H_s. It is obvious that as the still water level increases, the amplification coefficient is decreasing.

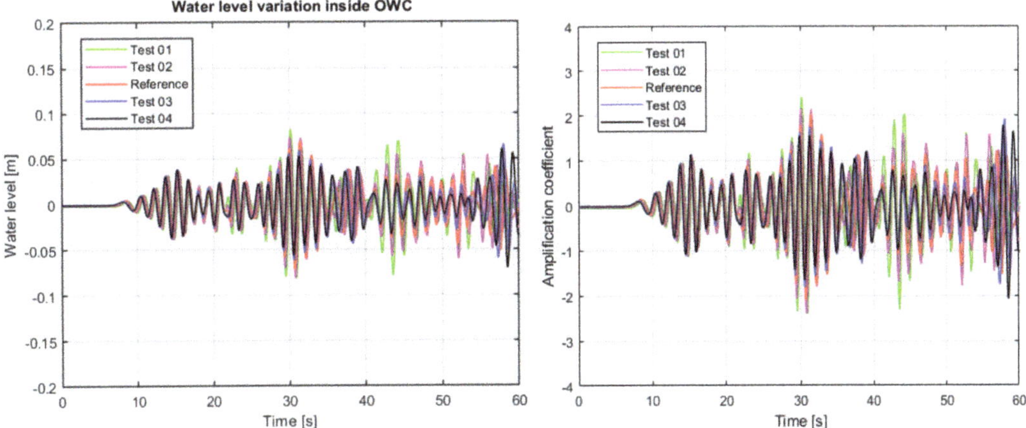

Figure 24. Comparison of the water level (**left** panel) and amplification coefficient (**right** panel) inside the OWC chamber for the Test_01 to Test_04 and Reference test.

On the other hand, increasing H_s leads to an increase in the water oscillation inside the OWC chamber, as shown in Figure 25—left panel. Nevertheless, the influence of H_s on the amplification coefficient which is derived from the division of the free surface elevation with the corresponding H_s is opposite (Figure 25—right panel). As the wave height increases the amplification coefficient decreases, as expected.

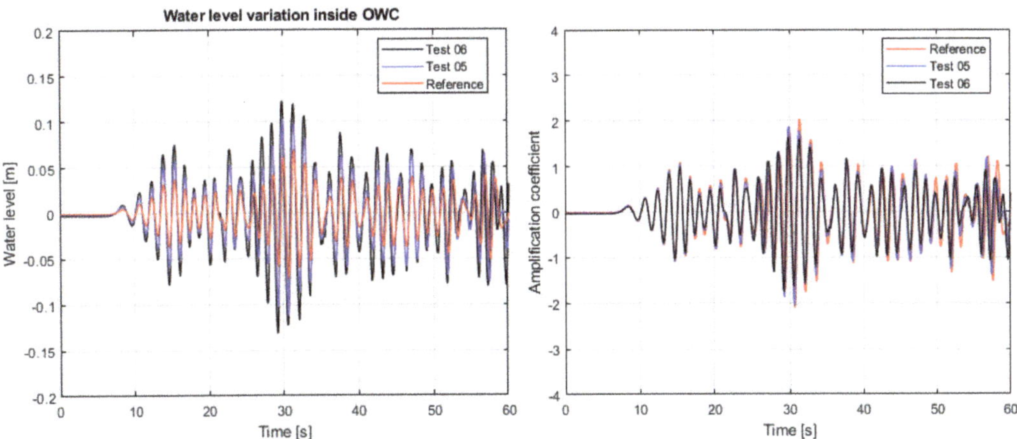

Figure 25. Comparison of the water level (**left** panel) and amplification coefficient (**right** panel) inside the OWC chamber for the Test_05 and Test_06 and Reference test.

6. Conclusions

Breakwater-integrated wave energy converters for harbour defence are being investigated in scale model tests, numerical models and full scale demonstrators for over 20 years now. Results to date show that, for most cases, the efficiency of the non-conventional breakwaters in terms of hydraulic performance is similar or even better compared to that of conventional structures. With respect to the structural response/stability of the non-conventional breakwaters, there is no evidence to date of fundamental negative impacts affecting the overall stability of the structure that are specific to the integration. Both these aspects—i.e., no detrimental effect to the hydraulic performance and structural response/stability, thereby not affecting the primary function of the breakwater that is providing sheltered conditions for port operations to develop—were corroborated by the first findings in the present research. The preliminary research findings further demonstrate that the integration is viable in both rubble mound and vertical breakwaters.

As far as electricity production go, the research suggests that the efficiency of land installed wave energy devices in converting wave power into electricity highly depends on the range of hydrodynamic conditions—i.e., waves and tidal levels—at the site and their variability. In the present research, hybridisation of suitable Wave Energy Converter (WEC) technologies for being integrated into harbour defence structures is hypothesised as an adequate approach to harness the available wave energy resource most efficiently over a wide range of metocean conditions. The underlying breakthrough idea is that hybridisation enables to exploit the strengths of each stand-alone technology and help overcome individual limitations and weaknesses. By comparing said suitable technologies through a multi-criteria analysis considering cost-effectiveness, constructability, WEC level of maturity, scalability/modularity, maintenance, reliability, etc. it was found that the preferred concepts for a hybridisation-based WEC technology for use in multi-purpose breakwaters are the Oscillating Water Column (OWC) and the Overtopping Device (OTD), which use two well understood power generating technologies, respectively, an air turbine and a water turbine.

The proof of concept of the innovative hybrid-WEC (HWEC) devised in the present research is based on a composite modelling approach combining physical and numerical modelling. The prototype breakwater and environmental conditions in the physical and numerical models are based upon on the rubble-mound cross-section proposed for the extension of the north breakwater of the Port of Leixões in Portugal. While the results of the physical model testing constructed to a scale of 1:50 are more extensively discussed in other papers (see e.g., [12,50,51]), this paper discusses the two-dimensional numerical modelling of the HWEC using the Computational Fluid Dynamics (CFD) software ANSYS Fluent.

The rubble-mound breakwater-integrated HWEC tested with ANSYS Fluent replicated the physical model tests performed by Cabral et al. [50]. Despite the non-conventional geometry of the HWEC device and complex wave-structure interactions, the numerical simulations showed that the ANSYS Fluent software is able to correctly reproduce the wave field in front of the breakwater, as well as the wave overtopping into the reservoirs and the water level oscillation on the OWC chamber, as measured in the laboratory. Once validated, the CFD model was used to test different geometrical configurations of the OWC entrance and identify the most effective geometry from a number of alternatives. The results obtained indicate that the OWC entrance plays a significant role in the overtopping discharges, since the tested alternatives imposed modifications in the front ramp of the OTD device. Impacts on the water oscillation inside the OWC chamber were also assessed relevant in the numerical modelling, similarly to what was concluded from the experimental testing. In fact, the optimization of the geometry of the HWEC with respect to one of the wave energy conversion principles was found to have relevant negative impacts on the other. Hence, the selected geometry for the HWEC is the one leading to the highest efficiencies and overall energy production. It should be stressed that the hydraulic/pneumatic and overall efficiencies reached 44% and 27%, respectively in the experimental testing, which shows the interest and the potential of this innovative hybrid-based technology.

Finally, additional numerical simulations with the most effective geometry were performed in order to extend the physical model testing programme to more water levels and wave conditions. These simulations confirmed the effect of the water level and wave conditions on both the overtopping discharges and water oscillations inside the OWC chamber. In the next phase of the breakwater-integrated HWEC development, three-dimensional numerical simulations will be carried out to better characterise and enhance the efficiency of the hybrid-based technology in converting wave power in electricity, including an estimation of the hydro- and aerodynamic coefficients for the design of the turbines. Moreover, a new experimental testing campaign will be performed to characterise wave loading on the structural elements of the HWEC, as well as to allow a higher resolution assessment of the impact of this technology on the rubble-mound breakwater. The new data and insights will be important to obtain an accurate estimation of the Levelised Cost of Energy (LCoE) of the HWEC technology.

The first findings in the present research further indicate that, for the same multipurpose breakwater and compared to the stand-alone WEC devices, the hybridisation-based WEC technology is a credible approach to reduce the LCoE.

Author Contributions: CFD-model set-up and simulations were done by IMDC (T.I.K.; E.D.L.; L.d.N.). The conceptualisation and execution of the physical model testing programme was of the responsibility of FEUP (T.C.-C.; P.R.-S.; F.T.-P.). The post-processing of the composite modelling approach results was done by IMDC (T.I.K.; E.D.L.; L.d.N.). The funding acquisition was the responsibility of the senior researchers of the R&D project (L.d.N.; F.T.-P.; P.R.-S.). All authors have read and agreed to the published version of the manuscript.

Funding: This research was funded by the OCEANERA-NET Second Joint Call 2016 project SE@PORTS—Sustainable Energy at sea PORTS (proposal ID 325), with the reference HBC.2016.0323 under the frame of Agentschap Innoveren en Ondernemen (VLAIO), as well as by the OCEANERA-NET Second Joint Call 2019 project WEC4PORTS—a hybrid Wave Energy Converter for PORTS (proposal ID 5111), with the references HBC.2019.2443, under the frame of Agentschap Innoveren en Ondernemen (VLAIO), and OCEAN-ERA/0004/2019, under the frame of FCT—Fundação para a Ciência e a Tecnologia.

Acknowledgments: The authors wish to acknowledge the remainder partners of the SE@Ports consortium, namely INEGI (project Coordinator), IH Cantabria, PLOCAN and Fórum Oceano. Furthermore, the authors are in debt to LNEC (*Laboratório Nacional de Engenharia Civil*), which lent the Antifer units used to materialise the armour layer of the reference breakwater model, to INEGI, who produced the HWEC physical model, and to the Port Authority of Douro, Leixões, and Viana do Castelo, for all the information provided to characterise the case study. João Henriques (from *Instituto Superior Técnico*) provided important contributions to the testing and control of the PTO system used in the OWC module.

Conflicts of Interest: The authors declare no conflict of interest. The funders had no role in the design of the study; in the collection, analyses or interpretation of data; in the writing of the manuscript, or in the decision to publish the results.

References

1. Barstow, S.; Mørk, G.; Mollison, D.; Cruz, J. The wave energy resource. In *Ocean Wave Energy*; Springer: Berlin/Heidelberg, Germany, 2008; pp. 93–132.
2. Greaves, D.; Iglesias, G. *Wave and Tidal Energy*; Wiley Online Library: Hoboken, NJ, USA, 2018; ISBN 9781119014447. Online ISBN 9781119014492. [CrossRef]
3. Magagna, D.; Uihlein, A. *2014 JRC Ocean Energy Status Report*; European Commission Joint Research Centre: Ispra, Italy, 2015.
4. Torre-Enciso, Y.; Ortubia, I.; De Aguileta, L.L.; Marqués, J. Mutriku wave power plant: From the thinking out to the reality. In Proceedings of the 8th European Wave and Tidal Energy Conference, Uppsala, Sweden, 7–10 September 2009; Volume 710.
5. Arena, F.; Romolo, A.; Malara, G.; Fiamma, V.; Laface, V. The first worldwide application at full-scale of the REWEC3 device in the Port of Civitavecchia: Initial energetic performances. In *Progress in Renewable Energies Offshore, Proceedings of the 2nd International Conference on Renewable Energies Offshore (RENEW2016), Lisbon, Portugal, 24–26 October 2016*; Taylor & Francis Group: London, UK, 2016; p. 303.
6. Contestabile, P.; Crispino, G.; Di Lauro, E.; Ferrante, V.; Gisonni, C.; Vicinanza, D. Overtopping breakwater for wave Energy Conversion: Review of state of art, recent advancements and what lies ahead. *Renew. Energy* **2020**, *147*, 705–718. [CrossRef]

7. Vicinanza, D.; Contestabile, P.; Nørgaard, J.Q.H.; Andersen, T.L. Innovative rubble mound breakwaters for overtopping wave energy conversion. *Coast. Eng.* **2014**, *88*, 154–170. [CrossRef]

8. Mustapa, M.A.; Yaakob, O.B.; Ahmed, Y.M.; Rheem, C.K.; Koh, K.K.; Adnan, F.A. Wave energy device and breakwater integration: A review. *Renew. Sustain. Energy Rev.* **2017**, *77*, 43–58. [CrossRef]

9. Falcão, A.F.; Henriques, J.C. Oscillating-water-column wave energy converters and air turbines: A review. *Renew. Energy* **2016**, *85*, 1391–1424. [CrossRef]

10. Vicinanza, D.; Lauro, E.D.; Contestabile, P.; Gisonni, C.; Lara, J.L.; Losada, I.J. Review of innovative harbor breakwaters for wave-energy conversion. *J. Waterw. Port Coast. Ocean Eng.* **2019**, *145*, 03119001. [CrossRef]

11. Das Neves, L.; Samadov, Z.; Di Lauro, E.; Delecluyse, K.; Haerens, P. The integration of a hybrid Wave Energy Converter in port breakwaters. In Proceedings of the 13th European Wave and Tidal Energy Conference, Naples, Italy, 1–6 September 2019.

12. Rosa-Santos, P.; Taveira-Pinto, F.; Clemente, D.; Cabral, T.; Fiorentin, F.; Belga, F.; Morais, T. Experimental study of a hybrid wave energy converter integrated in a harbor breakwater. *J. Mar. Sci. Eng.* **2019**, *7*, 33. [CrossRef]

13. Takahashi, S. *A Study on Design of a Wave Power Extracting Caisson Breakwater*; Wave Power Laboratory, Port and Harbour Research Institute: Negase, Japan, 1988.

14. Goda, Y. Random Seas and Maritime Structures. In *Advanced Series on Ocean Engineering*; World Scientific Publishing Company: Singapore, 2000; p. 15.

15. Müller, G.U.; Whittaker, T.J.T. An investigation of breaking wave pressures on inclined walls. *Ocean Eng.* **1993**, *20*, 349–358. [CrossRef]

16. Muller, G.; Whittaker, T.J.T. Field Measurements of Breaking Wave Loads on a Shoreline Wave Power Station. *Proc. Inst. Civ. Eng. Water Marit. Energy* **1995**, *112*, 187–197. [CrossRef]

17. Jayakumar, V.S. Wave Force on Oscillating Water Column Type Wave Energy Caisson: An Experiment Study. Ph.D. Thesis, Department of Ocean Engineering, Indian Institute of Technology, Madras, India, 1994.

18. Neumann, F.; Sarmento, A.J.N.A. OWC-caisson economy and its dependency on breaking wave design loads. In *The Eleventh International Offshore and Polar Engineering Conference*; International Society of Offshore and Polar Engineers: Mountain View, CA, USA, 2001.

19. Hull, P.; Müller, G. An investigation of breaker heights, shapes and pressures. *Ocean Eng.* **2002**, *29*, 59–79. [CrossRef]

20. Thiruvenkatasamy, K.; Neelamani, S.; Sato, M. Nonbreaking wave forces on multiresonant oscillating water column wave power caisson breakwater. *J. Waterw. Port Coast. Ocean Eng.* **2005**, *131*, 77–84. [CrossRef]

21. Patterson, C.; Dunsire, R.; Hillier, S. Development of wave energy breakwater at Siadar, Isle of Lewis. In *Coasts, Marine Structures and Breakwaters: Adapting to Change, Proceedings of the 9th International Conference Organised by the Institution of Civil Engineers, Edinburgh, UK, 16–18 September 2009*; Thomas Telford Ltd.: London, UK, 2010; pp. 1–738.

22. Huang, Y.; Shi, H.; Liu, D.; Liu, Z. Study on the breakwater caisson as oscillating water column facility. *J. Ocean Univ. China* **2010**, *9*, 244–250. [CrossRef]

23. Liu, Y.; Shi, H.; Liu, Z.; Ma, Z. Experiment study on a new designed OWC caisson breakwater. In Proceedings of the 2011 Asia-Pacific Power and Energy Engineering Conference, Wuhan, China, 25–28 March 2011; pp. 1–5.

24. Kuo, Y.S.; Lin, C.S.; Chung, C.Y.; Wang, Y.K. Wave loading distribution of oscillating water column caisson breakwaters under non-breaking wave forces. *J. Mar. Sci. Technol.* **2014**, *23*, 78–87.

25. Viviano, A.; Naty, S.; Foti, E.; Bruce, T.; Allsop, W.; Vicinanza, D. Large-scale experiments on the behaviour of a generalised Oscillating Water Column under random waves. *Renew. Energy* **2016**, *99*, 875–887. [CrossRef]

26. Naty, S.; Viviano, A.; Foti, E. Wave energy exploitation system integrated in the coastal structure of a Mediterranean port. *Sustainability* **2016**, *8*, 1342. [CrossRef]

27. Sainflou, G. Essai sur les digues maritimes verticales. *Ann. Ponts Chaussées* **1928**, *98*, 5–48.

28. Ashlin, S.J.; Sundar, V.; Sannasiraj, S.A. Pressures and forces on an oscillating water column–type wave energy caisson breakwater. *J. Waterw. Port Coast. Ocean Eng.* **2017**, *143*, 04017020. [CrossRef]

29. Pawitan, K.A.; Dimakopoulos, A.S.; Vicinanza, D.; Allsop, W.; Bruce, T. A loading model for an OWC caisson based upon large-scale measurements. *Coast. Eng.* **2019**, *145*, 1–20. [CrossRef]

30. Allsop, W.; Bruce, T.; Alderson, J.; Ferrante, V.; Russo, V.; Vicinanza, D.; Kudella, M. Large scale test on a generalised oscillating water column wave energy converter. In Proceedings of the HYDRALAB IV Joint User Meeting, Lisbon, Portugal, 2–4 July 2014.

31. Zanuttigh, B.; Margheritini, L.; Gambles, L.; Martinelli, L. Analysis of wave reflection from wave energy converters installed as breakwaters in harbour. In Proceedings of the European Wave and Tidal Energy Conference (EWTEC), Uppsala, Sweden, 7–10 September 2009.

32. US Army Corps of Engineers. *Coastal Engineering Manual. Engineer Manual 1110-2-1100*; US Army Corps of Engineers: Washington, DC, USA, 2002.

33. Simonetti, I.; Cappietti, L. Hydraulic performance of oscillating water column structures as anti-reflection devices to reduce harbour agitation. *Coast. Eng.* **2020**. [CrossRef]

34. Contestabile, P.; Ferrante, V.; Di Lauro, E.; Vicinanza, D. Full-scale prototype of an overtopping breakwater for wave energy conversion. *Coast. Eng. Proc.* **2017**, *1*, 12. [CrossRef]

35. Margheritini, L.; Vicinanza, D.; Frigaard, P. SSG wave energy converter: Design, reliability and hydraulic performance of an innovative overtopping device. *Renew. Energy* **2009**, *34*, 1371–1380. [CrossRef]

36. Margheritini, L.; Stratigaki, V.; Troch, P. Geometry optimization of an overtopping wave energy device implemented into the new breakwater of the Hanstholm port expansion. In *The Twenty-Second International Offshore and Polar Engineering Conference*; International Society of Offshore and Polar Engineers: Mountain View, CA, USA, 2012.
37. Buccino, M.; Vicinanza, D.; Salerno, D.; Banfi, D.; Calabrese, M. Nature and magnitude of wave loadings at Seawave Slot-cone Generators. *Ocean Eng.* **2015**, *95*, 34–58. [CrossRef]
38. Contestabile, P.; Iuppa, C.; Di Lauro, E.; Cavallaro, L.; Andersen, T.L.; Vicinanza, D. Wave loadings acting on innovative rubble-mound breakwater for overtopping wave energy conversion. *Coast. Eng.* **2017**, *122*, 60–74. [CrossRef]
39. Buccino, M.; Dentale, F.; Salerno, D.; Contestabile, P.; Calabrese, M. The use of CFD in the analysis of wave loadings acting on Seawave slot-cone generators. *Sustainability* **2016**, *8*, 1255. [CrossRef]
40. Di Lauro, E.; Lara, J.L.; Maza, M.; Losada, I.J.; Contestabile, P.; Vicinanza, D. Stability analysis of a non-conventional breakwater for wave energy conversion. *Coast. Eng.* **2019**, *145*, 36–52. [CrossRef]
41. Di Lauro, E.; Maza, M.; Lara, J.L.; Losada, I.J.; Contestabile, P.; Vicinanza, D. Advantages of an innovative vertical breakwater with an overtopping wave energy converter. *Coast. Eng.* **2020**, *159*, 103713. [CrossRef]
42. Palma, G.; Contestabile, P.; Zanuttigh, B.; Formentin, S.M.; Vicinanza, D. Integrated assessment of the hydraulic and structural performance of the OBREC device in the Gulf of Naples, Italy. *Appl. Ocean Res.* **2020**, *101*, 102217. [CrossRef]
43. Iuppa, C.; Contestabile, P.; Cavallaro, L.; Foti, E.; Vicinanza, D. Hydraulic performance of an innovative breakwater for overtopping wave energy conversion. *Sustainability* **2016**, *8*, 1226. [CrossRef]
44. EurOtop. *Manual on Wave Overtopping of Sea Defences and Related Structures. An Overtopping Manual Largely Based on European Research, but for Worldwide Application*; Van der Meer, J.W., Allsop, N.W.H., Bruce, T., De Rouck, J., Kortenhaus, A., Pullen, T., Schüttrumpf, H., Troch, P., Zanuttigh, B., Eds.; Available online: www.overtopping-manual.com (accessed on 15 February 2021).
45. Iuppa, C.; Cavallaro, L.; Musumeci, R.E.; Vicinanza, D.; Foti, E. Empirical overtopping volume statistics at an OBREC. *Coast. Eng.* **2019**, *152*, 103524. [CrossRef]
46. Cavallaro, L.; Iuppa, C.; Castiglione, F.; Musumeci, R.E.; Foti, E. A Simple Model to Assess the Performance of an Overtopping Wave Energy Converter Embedded in a Port Breakwater. *J. Mar. Sci. Eng.* **2020**, *8*, 858. [CrossRef]
47. Cappietti, L.; Simonetti, I.; Penchev, V.; Penchev, P. Laboratory tests on an original wave energy converter combining oscillating water column and overtopping devices. In Proceedings of the 3rd International Conference on Renewable Energies Offshore (RENEW-2018), Lisbon, Portugal, 8–10 October 2018.
48. Lara, J.; de Eulate, M.Á.; Di Paolo, B.; Rodríguez, B.; Guanche, R.; Álvarez, A.; Mendoza, A.; Iturrioz, A.; Blanco, D.; Di Lauro, E. Physical and Numerical Modeling of an Innovative Vertical Breakwater for Sustainable Power Generation in Ports: SE@ PORTS Project. *Coast. Struct.* **2019**, 158–168. [CrossRef]
49. Frostick, L.E.; McLelland, S.J.; Mercer, T.G. (Eds.) *Users Guide to Physical Modelling and Experimentation: Experience of the HYDRALAB Network*; CRC Press: Boca Raton, FL, USA, 2011.
50. Cabral, T.; Clemente, D.; Rosa-Santos, P.; Taveira-Pinto, F.; Morais, T.; Belga, F.; Cestaro, H. Performance Assessment of a Hybrid Wave Energy Converter Integrated into a Harbor Breakwater. *Energies* **2020**, *13*, 236. [CrossRef]
51. Cabral, T.; Clemente, D.; Rosa-Santos, P.; Taveira-Pinto, F.; Morais, T.; Cestaro, H. *Evaluation of the Annual Electricity Production of a Hybrid Breakwater-Integrated Wave Energy Converter*; Elsevier: Amsterdam, The Netherlands, 2020; Volume 213, p. 118845. [CrossRef]
52. ANSYS. *ANSYS Fluent Theory Guide*; Release 15.0; ANSYS: Canonsburg, PA, USA, 2013; p. 780.

Journal of
*Marine Science
and Engineering*

Article

Hydraulic and Structural Assessment of a Rubble-Mound Breakwater with a Hybrid Wave Energy Converter

Daniel Clemente [1,2,*], Tomás Calheiros-Cabral [1,2], Paulo Rosa-Santos [1,2] and Francisco Taveira-Pinto [1,2]

[1] Interdisciplinary Centre of Marine and Environmental Research, University of Porto, Terminal de Cruzeiros do Porto de Leixões, Avenida General Norton de Matos, s/n, 4450-208 Matosinhos, Portugal; tcabral@fe.up.pt (T.C.-C.); pjrsantos@fe.up.pt (P.R.-S.); fpinto@fe.up.pt (F.T.-P.)

[2] Department of Civil Engineering, Faculty of Engineering, University of Porto, Rua Dr. Roberto Frias, s/n, 4200-465 Porto, Portugal

* Correspondence: up201009043@edu.fe.up.pt

Citation: Clemente, D.; Calheiros-Cabral, T.; Rosa-Santos, P.; Taveira-Pinto, F. Hydraulic and Structural Assessment of a Rubble-Mound Breakwater with a Hybrid Wave Energy Converter. *J. Mar. Sci. Eng.* **2021**, *9*, 922. https://doi.org/10.3390/jmse9090922

Academic Editors: Giuseppe Roberto Tomasicchio and Eva LOUKOGEORGAKI

Received: 23 July 2021
Accepted: 22 August 2021
Published: 25 August 2021

Publisher's Note: MDPI stays neutral with regard to jurisdictional claims in published maps and institutional affiliations.

Abstract: Seaports' breakwaters serve as important infrastructures capable of sheltering ships, facilities, and harbour personnel from severe wave climate. Given their exposure to ocean waves and port authorities' increasing awareness towards sustainability, it is important to develop and assess wave energy conversion technologies suitable of being integrated into seaport breakwaters. To fulfil this goal whilst ensuring adequate sheltering conditions, this paper describes the performance and stability analysis of the armour layer and toe berm of a 1/50 geometric scale model of the north breakwater extension project, intended for the Port of Leixões, with an integrated hybrid wave energy converter. This novel hybrid concept combines an oscillating water column and an overtopping device. The breakwater was also studied without the hybrid wave energy device as to enable a thorough comparison between both solutions regarding structural stability, safety, and overtopping performance. The results point towards a considerable reduction in the overtopping volumes through the integration of the hybrid technology by an average value of 50%, while the stability analysis suggests that the toe berm of the breakwater is not significantly affected by the hybrid device, leading to acceptable safety levels. Even so, some block displacements were observed, and the attained stability numbers were slightly above the recommended thresholds from the literature. It is also shown that traditional damage assessment parameters should be applied with care when non-conventional structures are analysed, such as rubble-mound breakwaters with integrated wave energy converters.

Keywords: physical modelling; wave energy; breakwaters; safety; overtopping; stability

1. Introduction

Seaports are important maritime infrastructures responsible for the development of regional economies and the global transport trade. Yet, this has some consequences, as pollution is a source of great concern [1] in terms of noise [2,3], greenhouse gas (GHG) emissions [4], and public health [5]. They can also suffer from inherent environmental consequences, given their coastal exposure and vulnerability to sea-level rise, storms, and in some cases, flooding [6,7]. This has drawn the attention and concern of port authorities and stakeholders, who have started to identify and implement new and more sustainable policies within seaports [8–10] in order to mitigate the negative impacts of seaport activities.

One of the most promising options being considered by port authorities relies upon the integration of wave energy converters (WECs) into breakwaters [11–13], which represent the main port structures responsible for sheltering the inner areas of seaports and berthing ships from overtopping and severe wave climate [14,15]. This approach presents an opportunity to supply seaports with clean and environmentally sustainable energy by harnessing a directly available resource, given ports' exposure to ocean waves, with a high global potential [16,17]. However, it is crucial that the main functions of breakwaters are assured, as well as their structural stability. The functionality of rubble-mound breakwaters, for instance, can be compromised due to several occurrences. These include the displacement of blocks by plunging waves, liftouts by uprush and downrush, slides of the armour layers, and/or failure due to special wave trains and toe berm erosion, among others [14]. Several formulae and methods have been developed for the design and stability analysis (e.g., based on the stability number, N_s) [18–21]. Numerous studies have also been carried out on the damage assessment (e.g., based on the damage number, N_{od}), namely, of the toe berm for rubble-mound breakwaters [22,23]. Studies on the armour layer are more focused on the stability number, N_S, which is related to the determination of the armour block's weight [24] through empirical formulae [25,26]. Most studies are mainly aimed at assessing damage on the structure during storms, as this is the main source of damage during a breakwater's usable life [27]. Lastly, though this damage assessment methodology can be used for breakwater integrated WECs, such as overtopping devices (OWEC) [28] and oscillating water columns (OWC) [29], it is crucial to ensure that breakwaters are also able to cope with wave run-up and overtopping, in order to protect innermost infrastructures of the harbour and berthing ships. Several projects have been carried out to attain information and create databases to serve as design references, such as the CLASH [30–32] and the VOWS [33] projects. To these we add several experimental [30,34] and numerical studies [35,36], as well as field measurements [37,38] and recommendations from various sources on overtopping limits, such as the EurOtop Manual [39,40]. In summary, not only is it pertinent to select and study an adequate WEC technology from a wide range of solutions, but it is equally relevant to ensure that its integration into a port structure, such as a breakwater, does not compromise its structural stability or functionality.

This paper discusses the integration of a combined OWEC–OWC wave energy converter concept into a rubble-mound breakwater, with particular focus on the consequential influence in terms of structural stability and functionality of the sheltering structure. Section 2 of this paper describes the experimental study carried out with a reduced scale physical model of the breakwater and WEC, the equipment that was utilized, the characteristics of the experimental facility, and the case study used as reference. Section 3 presents and discusses the most relevant results under the scope of this paper. The overtopping volumes, stability numbers, and damage numbers for a series of severe sea-state tests are compared with and without the integration of the hybrid WEC. Finally, Section 4 summarizes the main conclusions of the experimental study and discusses the main consequences of the proposed WEC solution from a structural perspective.

2. Experimental Study of the Hybrid OWEC–OWC Device

2.1. Case Study and Test Plan

The integration of the hybrid WEC concept has been considered for the case study of the Port of Leixões, in the north-western coast of Portugal (41°11′ N and 8°42′ W), 4 km north of the river Douro's mouth. This seaport was selected due to its economic importance at a regional and national level, the highly energetic wave climate to which it is subjected, and the development strategy being followed by the Port Authority of Douro, Leixões, and Viana do Castelo (APDL). Amongst other expansion and infrastructural improvement plans, the port authority is currently considering an extension of the north breakwater by 300 m [41]. The main goal is to allow for ships of greater length/draught to safely enter the port during rough wave conditions and to provide better tranquillity conditions within the harbour basin.

In fact, facing the Atlantic Ocean, the Port of Leixões (Figure 1) is very susceptible to highly energetic sea states. During storms, significant wave heights have been commonly recorded at 8 m or more, with a maximum registered value slightly above 10 m, in 2014 [41]. These data were obtained from offshore wave buoy measurements (deep water conditions). Hence, the north breakwater serves as one of the key protection structures of the innermost areas of the port, alongside the south breakwater.

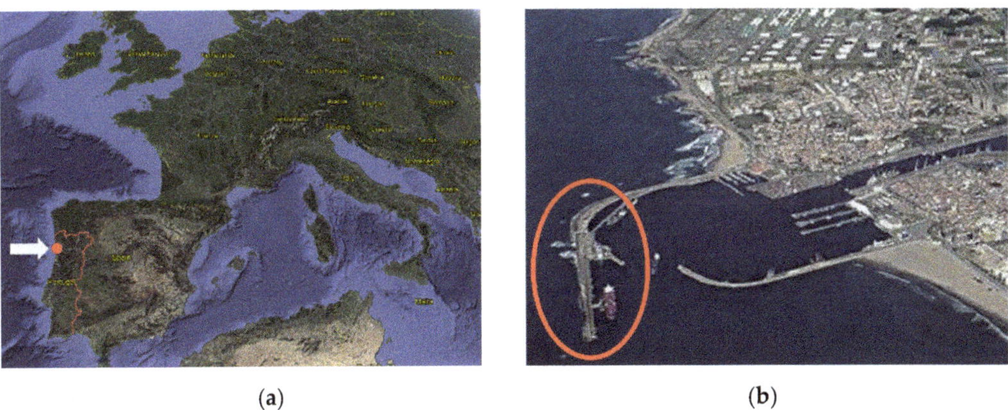

Figure 1. Port of Leixões: (**a**) location in the Iberian Peninsula and (**b**) aerial view of the port and its breakwaters (north breakwater highlighted).

Current plans for the improvement of the port accessibility mention an extension of the north breakwater by about 300 m, at an angle of 20° towards the west regarding the existing structure's alignment (Figure 2a). This presents a unique opportunity to assess the potential of an integrated hybrid WEC (HWEC) into the extended breakwater. The proposed solution entails a rubble-mound breakwater with an armour layer composed of a double layer of Antifer blocks on the seaward side and a single layer on the inwards side (Figure 2c). A 20-metre-long section was considered for this study, with the HWEC being integrated at the centre of the breakwater (Figure 2b) at a water depth of circa 17.5 m (reference to the mean sea level or M.S.L.).

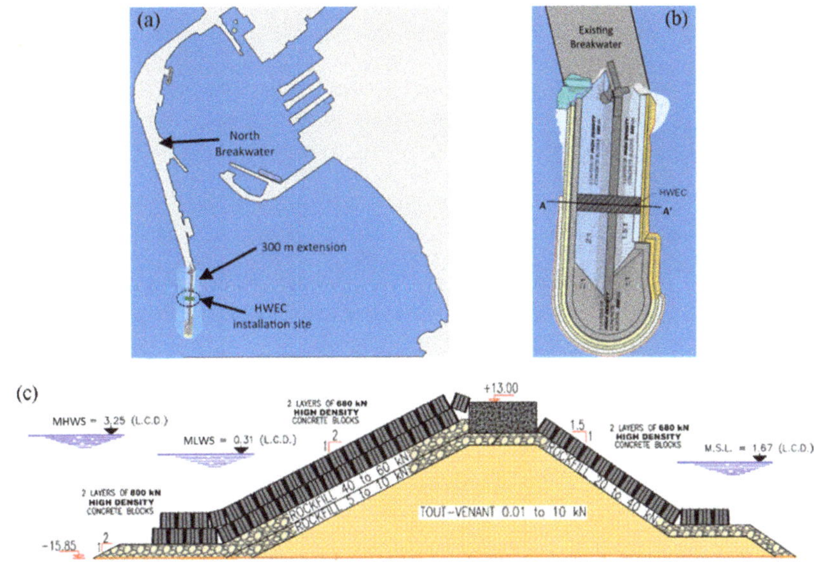

Figure 2. (**a**) General layout of Port of Leixões with the breakwater extension and the HWEC; (**b**) provisional layout of the extension of the breakwater; (**c**) cross-section AA' with the three water levels represented, reproduced with permission from [42]. Copyright 2020, Elsevier (cf. Table 1).

Table 1. Sea states tested for the assessment of overtopping flow and structural stability (prototype values).

Sea State	H_S (m)	T_P (s)	Water Level
SS1	6.0	13	
SS2	7.7	16	MLWS
SSMax	8.0	16	(+0.31 m L.C.D)
SS1	6.0	13	
SS2	7.7	16	MHWS
SSMax	9.1	16	(+3.25 m L.C.D)

The experimental study with the physical model, which is later described, encompassed a wide range of wave conditions based on data from the case study area, as to study the HWEC's energy conversion performance under operational conditions [42,43]. Nevertheless, in order to reproduce extreme events and assess the impact that they can have on a future breakwater extension, an additional set of extreme wave conditions were chosen. This took into consideration the water depth and the wavemaker's physical limitations. Table 1 characterizes the extreme sea states tested for the mean low water springs (MLWS) and mean high water springs (MHWS).

Standard procedures for reproduction of irregular sea states originated from the North Atlantic Ocean involve the definition of a JONSWAP spectrum, with a peak enhancement factor of 3.3. To ensure an adequate reproduction of the sea states and enable a deterministic result comparison, a temporal sequence of circa 1200 waves was used for each test [44]. However, it should be noted that the most severe wave conditions (highest significant wave height, H_s, and peak wave period, T_p) were defined based on the physical limitations imposed by the wavemaker system, which operated with an active absorption module to mitigate wave reflection. Wave breaking phenomena in the foreshore, at the front of the rubble-mound breakwater, was also taken into account.

2.2. OWEC–OWC Concept and Physical Model

The hybrid WEC concept resulted from the combination of an overtopping device (OWEC) [45–47] with an oscillating water column (OWC) [29,48,49] system [50]. For a single module, it is expected that wave energy conversion be achieved through a dual-mode operation, which should yield a greater performance than that of standalone variants of the two original concepts, as it was shown in [42,43,51]. To accommodate the different types of WEC whilst mitigating modifications of the preliminary design of the breakwater structure, several adaptations were introduced.

The hybrid WEC's dimensions and cross-section are compatible with the original structural design of the rubble-mound breakwater, from the toe berm to the crest. The OWC chamber was integrated amidst the reservoirs, intercepting them at a central position and connecting to an upper section where the power take-off (PTO) would be located. However, the OWC is isolated from the OWEC component. In the interior of the chamber, there is an air volume trapped between the PTO (likely a turbine) and the seawater that comes from the intake of the OWC. The passing of waves induces fluctuations on the free surface of the water inside the chamber, which, in turn, generates air pressure differentials that drive the PTO's energy conversion process. As for the OWEC, several reservoirs allow for the intake and accumulation of overtopping volumes. The different levels allow for various combinations of sea level and wave height to be harnessed. The ramp slope is similar to that of the original breakwater structure design, while the reservoirs denote a small inclination towards the interior of the WEC, in order to promote the flow of seawater into the penstocks associated to the hydraulic turbines (in the case of the physical model, pipes), as seen in Figure 3. Low-head turbines convert the available potential energy (head and stored water) into electric energy, similar to the process observed in hydroelectric plants. Lastly, the shape of the device's crest was selected as to serve as a barrier, which not only mitigates overtopping of the structure, but also promotes the accumulation of water inside the OWEC's reservoirs. Even so, as described above, it was important to measure the overtopping volumes, especially those related to the hybrid WEC's crest. The resulting setup is presented in Figure 3, which depicts the connections between the physical model's OWEC reservoirs and the corresponding auxiliary reservoirs (ARs).

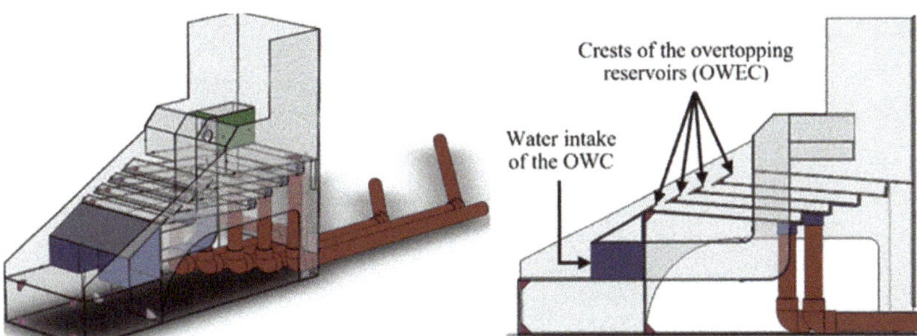

Figure 3. Hybrid WEC model scheme for construction in SolidWorks®. Reproduced with permission from [42]. Copyright 2020, Elsevier.

2.3. Facilities, Materials, and Equipment

For the purpose of the physical modelling stage, a 3D reproduction of the hybrid OWEC–OWC device at a geometric scale of 1:50, following on the Froude similitude criteria, was considered (Figure 4). The selected domain was a channel that extended 14.30 m in length and 0.84 m in width, within the wave basin at the Hydraulics Laboratory of the Hydraulics, Water Resources and Environment Division of the Faculty of Engineering of the University of Porto (FEUP), Portugal. The wave basin itself spans 28 m in length, 12 m in width, and 1.2 m in height and has a multi-element piston-type wave generation system (manufactured by HR Wallingford, UK) that is used for reproducing a wide variety of wave conditions, including regular waves and irregular long-crested waves. The system also includes a dynamic wave absorption system to compensate for wave re-reflections.

Figure 4. Frontal view of the OWEC–OWC model (**left**) and perspective of the physical model setup for the extreme wave tests (**right**).

Special care was taken during the setup of the physical model in order to mitigate laboratorial and scale effects, such as the wall effects, given the limited dimensions of the domain, the need to mitigate potential gaps in the setup of the blocks, and the interest in reproducing an armour layer section as wide as feasible wide a section as feasible of the armour layer. This would enable a more thorough and direct comparison with the adjacent HWEC section. The width of the domain also implied that a single piston of the wavemaker system would be used to reproduce the selected regular and irregular long-crested wave conditions.

The bathymetry in front of the structure was simplified and reproduced to represent the relevant wave transformation processes during the propagation from the wavemaker paddle to the model, analogously to wave propagation from the location where the wave climate was established to the device's location in the breakwater. For this purpose, a sandy slope with an angle of 0.63° to the horizontal, on an overall length of 12.65 m, was created. As the evolution of the bathymetric profile over time to track any possible scouring phenomena occurring near the toe berm had already been monitored during a previous phase of the project [52], and no relevant scouring was detected, only visual observation of the sea bed in front of the structure was undertaken in this study.

In terms of overtopping measurements, the volume of water that overtops the break-water crest, with or without the hybrid WEC integrated, was collected in an AR positioned behind the physical model. The width of the collecting channel was the same in both configurations (0.40 m at model scale). In more detail, the overtopping volumes were collected by the upmost reservoir of the physical model and drained into the AR, at a lower position, through a connecting pipe. Due to the confined space, the overtopping volumes would accumulate, over time, within the AR, where a resistive wave gauge is installed to measure the variation of the free surface of water (FSW). To avoid overflow of the AR, a hydraulic pump was placed inside it, alongside the wave gauge. Hence, once an upper threshold is reached, the pump begins to operate and stops only when a very low elevation limit is achieved, close to 0 m. Since the inner area of the reservoir, A_r, is known, the overtopping volume, ΔV, can be calculated, for an interval of time, Δt, by taking the water elevation variation, Δh, and multiplying it by A_r. The summation of the overtopping volumes, ΔV, gives the total overtopping volume over the duration of the test. As a complement, during testing, it was possible to account for the pumped-out volumes of water, since the pump's flow curve and working time are also known.

Five additional gauges were deployed in this study and separated into two groups: a single gauge next to the physical model and the remaining four at a distance of, at least, one wavelength from the wavemaker's piston paddle. These were used to measure the FSW directly in front of the physical model and to perform a reflection analysis of the generated waves, respectively. The reflection analysis, and hence, the determination of the incident wave conditions, was done using a script based on a development of the least square method proposed by [53].

The experimental setup is summarized in Figure 5, which also presents the ARs that were used to collect the water gathered by the hybrid WEC's reservoirs, namely, the overtopping component, which is related to wave energy conversion under operational conditions (for additional information, see [42,43]).

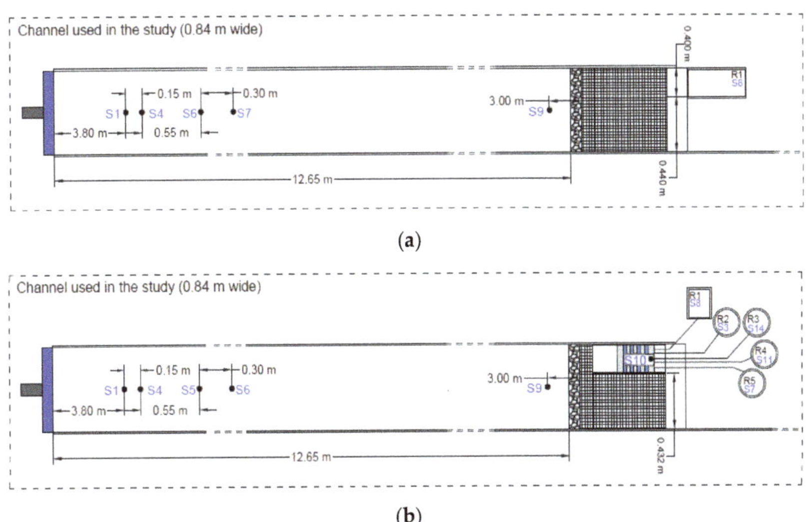

(a)

(b)

Figure 5. Experimental setup: (**a**) initial case, without WEC; (**b**) case with integrated WEC. For the extreme wave tests, which are the scope of this paper, only reservoir R1 was active.

3. Results

3.1. Structure's Functionality

Breakwaters' crest levels are a critical factor that should be carefully considered upon design. On one hand, crest heights may be required to be kept under a specific threshold, in order to minimize their visual impact. This is the case of the Port Leixões, as the extension of the north breakwater would introduce a new visual obstacle from the perspective of nearby residents and beach users. On the other hand, lower crest levels lead to more frequent overtopping occurrences, which can result in structural damage and longer periods of port terminal inoperativeness, should these exist. Therefore, the mean overtopping discharge should be studied for the most severe wave conditions that a breakwater is subjected to, as well as other more frequent conditions when overtopping might occur.

As explained previously, the overtopping flow over the crest of the structure is captured by a reservoir, on top of the device, connected to an AR on the back of the physical model. An example of a recorded time series is presented in Figure 6, where one can see the incrementing water level within the reservoir over time as well as the pump operation times.

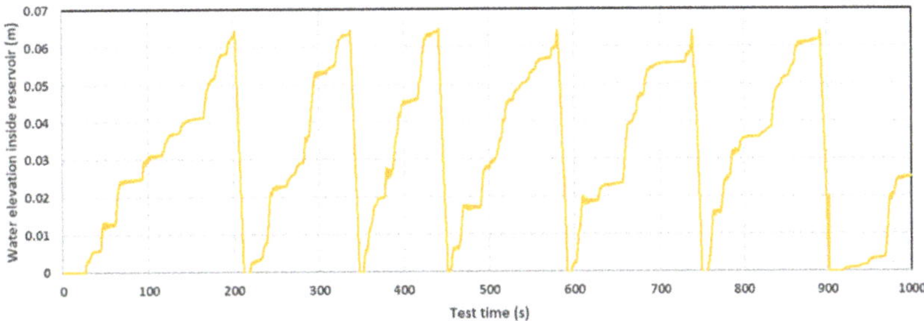

Figure 6. Time series of water level measurements inside an auxiliary reservoir (model values).

The analysis of the overtopping discharges was carried out having as reference the EurOtop manual [39], which presents two methods: one based on the mean value approach and another on the design and assessment approach, being the difference in the definition of two coefficients *a* and *b*. Because the former is recommended for predictions and comparisons with measurements, the mean value approach was employed. As such, the curve equation is defined as:

$$\frac{q}{\sqrt{gH_s^3}} = 0.09exp\left[-\left(-1.5\frac{R_c}{H_s\gamma_f\gamma_\beta}\right)^{1.3}\right] \tag{1}$$

where *q* is the mean overtopping discharge, *g* the acceleration of gravity, R_c the crest height relative to the mean sea level, H_s the spectral significant wave height, and γ_f and γ_β the roughness and oblique wave attack factors, respectively.

It is worth noting that, in Equation (1), γ_β and γ_f are given based on the results from the EU research program CLASH. A priori, γ_β is equal to 1, as in the study carried out, the waves do not act upon the breakwater structure at an oblique angle. However, the value of γ_f is worthy of discussion. From the CLASH results, it is recommended that, for Antifer blocks, a value of 0.50 should be used for this coefficient. Nevertheless, this value was attained for an Antifer block configuration that is far more disordered than that used in this study, where the blocks were aligned carefully both in the longitudinal and transversal directions (regular placement). This will become more perceivable in the following sub-section. There is also some data dispersion inherent to the CLASH study and other limitations in terms of transposition to the study that is the scope of this paper. As such, an adjustment of the γ_f coefficient was performed, aimed at approximating the relative overtopping rate curve (left-hand side of Equation (1)) to the experimental values, first for the case without the WEC, and aimed at minimizing the root mean square error (RMSE). Afterwards, the data for the case with the WEC integrated were considered (Figure 7).

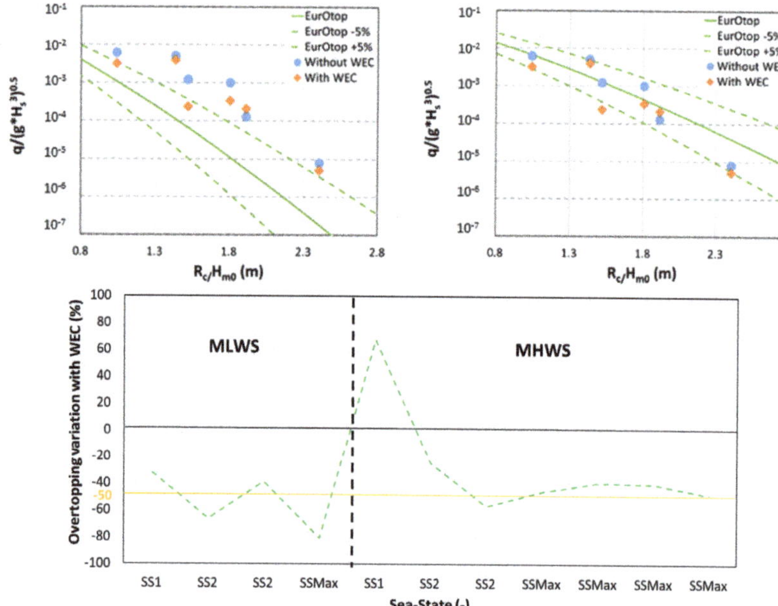

Figure 7. Relative overtopping rate curves following the mean value approach using a roughness coefficient of 0.50 (**left-hand side**, reference from EurOtop manual) and 0.77 (**right-hand side**). Overtopping discharge variation for each sea state (**bottom**) with the integration of the HWEC.

It was found that, for the case study of the Port of Leixões, a value of about 0.66 and 0.77 for the roughness coefficient would yield the best curve approximation for the experimental data, with and without the HWEC, respectively (see Figure 7, which includes 5% variability curves, in accordance to the EurOtop manual). These values were computed with the assistance of the non-linear generalized reduced gradient solver engine, which minimized the RMSE (1.07×10^{-5} and 1.11×10^{-5}, respectively). These γ_f values are larger than the recommended value of 0.50. This result is justified by the more regular arrangement of the blocks in the armour layer in comparison to the CLASH experiments. The Antifer arrangement in the Leixões case study creates a smoother surface, somewhat comparable to a ramp, thus leading to higher values of the relative overtopping rate.

The results from Figure 7 also demonstrate a small, yet important overall reduction in the relative overtopping rate with the introduction of the WEC. In fact, it was estimated that the HWEC enabled an average reduction of about 50% in the mean overtopping discharge for the considered test conditions (Figure 7), if SS1 for the MHWS is not considered. Even so, for this particular case, the attained variation was attributed to the usual uncertainty in the prediction of overtopping discharges [34], especially high for very low discharge data, which typically show a wider variability.

From a practical perspective, it is also pertinent to address the hypothetical use of the breakwater's inner side for different port related activities as well as the potential risks that are inherent to the occurrence of extreme sea states. Currently, the north breakwater of the Port of Leixões shelters from wave action an oil terminal, yet the extension project foresees, in essence, the improvement of the accessibility and navigation conditions within the harbour area. Hence, presently, no specific use of the extension is expected regarding road traffic, pedestrians, and alike activities, excepting those required for the construction and maintenance of the new structure. However, the situation may change in the future, as in the past, with the construction of the oil terminal of Leixões in the lee side of the existing north breakwater (Figure 1b). In that scenario, given the measured overtopping volumes and average discharges, the usage of interior areas of the north breakwater's extension would need to be severely restricted during rough environmental conditions. This is perceivable from Figure 8, which is based upon the limits recommended in the Coastal Engineering Manual [21].

Even by considering the thresholds provided by the Rock Manual [20] or the EurOtop Manual [39], access to pedestrians and vehicles would be deemed unsafe during the occurrence of extreme wave conditions (overtopping discharges exceed limits by, at least, two orders of magnitude). The berthing of small ships is also ill-advised for those conditions. The introduction of the HWEC into breakwater, albeit reducing the overall overtopping discharges, is insufficient to eliminate the need for additional measures to reduce the risks linked to overtopping. In addition, it will be important to adequately plan the construction process of the breakwater.

The aforementioned reflection analysis routine was employed for the estimation of the reflection coefficients, K_r (Figure 9). From the results, it is perceivable that the introduction of the HWEC leads to an overall reduction in the reflection coefficient for MLWS, while an increment is observed for MHWS. This could be justified by the structural design of the hybrid solution:

- For MLWS, the overtopping discharges are partially received by the HWEC's reservoirs. This exhibits a greater effect than that associated with reflections or energy dissipations by the original breakwater armour layer;
- For MHWS, significant volumes of overtopping discharges are still received by the reservoirs. However, given the higher waterline level, the overtopping flow easily reaches the upper section of the HWEC. The reflections upon the crest's vertical sections are responsible for the increment of K_r.

The order of magnitude does not vary considerably (0.28 to 0.42, approximately) and tends to stabilize beyond SS1. Additionally, although the significant wave height continues to increase, the peak wave period is conserved (16 s), which is in agreement with the evolution pattern of K_r.

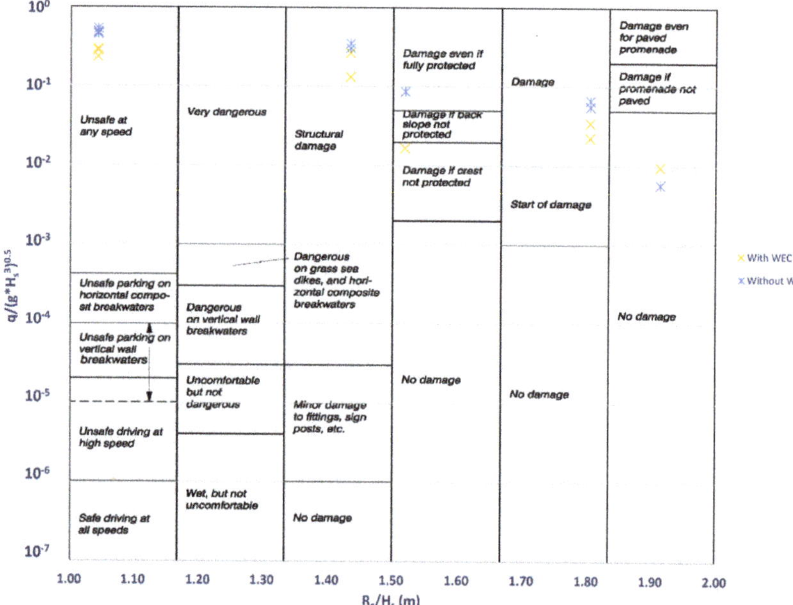

Figure 8. Critical values of average overtopping discharges versus experimental data results from the Leixões case study, adapted from [21].

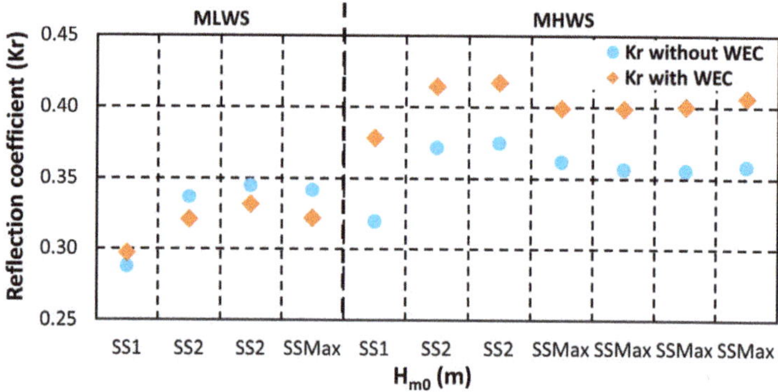

Figure 9. Reflection coefficients with and without the HWEC.

3.2. Structure's Stability

The analysis of the structural stability of the breakwater requires a comparison between the design solutions for the north breakwater extension with and without the integration of the hybrid OWEC–OWC device. It is crucial that the introduction of the WEC does not significantly weaken, considering a worst-case scenario, the structural stability of the rubble-mound breakwater, given its key importance as a sheltering structure of the innermost areas of the seaport. For the Port of Leixões case study, the armour layer and toe berm of the outer sections of the structure, composed of Antifer concrete blocks, are of particular interest. Any movements, rotations, or excessive damage require repositioning and/or replacement of those blocks. Therefore, it is crucial to identify sections susceptible to these occurrences, particularly for severe wave climate, and quantify the motions and number of blocks that have moved over the various tests.

A preliminary analysis of the evolution of the toe berm (light green blocks) and the outer amour layer (remaining blocks) of the breakwater is perceptible in Figure 10, for both the tests associated with MLWS and MHWS. It is noticeable that the outer armour layer has very small changes, both with and without the WEC; however, there are important changes that can be observed on the toe berm blocks. In the absence of the WEC, only three of these blocks were considerably displaced, and the number did not increase going from MLWS to MHWS tests. In contrast, when the WEC is integrated into the breakwater, the number of displaced blocks increases significantly. Moreover, unlike the previous case, there is an increment of the number of displaced blocks from the toe berm when comparing the final tests of MLWS and MHWS levels. The blocks are also shifted further away from the physical model.

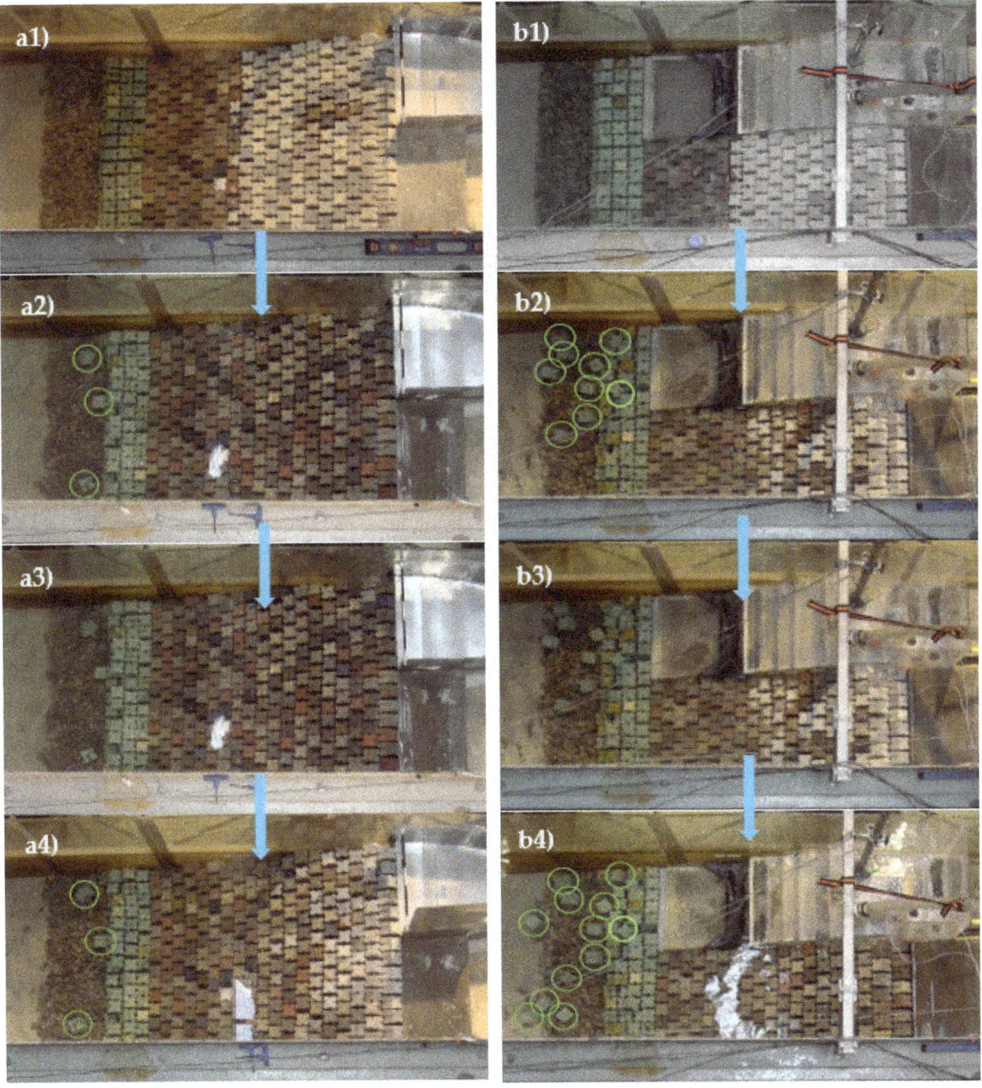

Figure 10. Evolution of the toe berm blocks and outer armour layer blocks, with (**a1–a4**) and without (**b1–b4**) the WEC: for MLWS, from initial disposition, (1), to final test, (2); for MHWS, from final test of MLWS, (3), to final test of the series, (4).

The stability of the outer armour layer and toe berm of the breakwater was analysed through the application of methods recommended in the literature [21]. An initial assessment involved the quantification of the damage number, N_{od}, defined as the number of displaced units within a breakwater strip with a width of D_n, and the stability number, N_s, a commonly used parameter to assess structural stability under wave attack [25]:

$$N_{od} = \frac{N_{dis}}{W/D_n} \tag{2}$$

$$N_s = \frac{H_s}{(\frac{\rho_s}{\rho_w} - 1)D_n} = (1.6 + 0.24\frac{h_b}{D_n})N_{od}^{0.15} \tag{3}$$

where N_{dis} represents the number of units displaced out of the armour layer or toe berm, W the width of the reference section, D_n the nominal diameter of the blocks, h_b the water depth at the top of the toe berm, and ρ_s and ρ_w the mass density of the blocks and water, respectively. Considering that Antifer blocks were used in this study, D_n was calculated as the equivalent cube length. Equation (3) refers to the stability of a toe berm formed by two layers of stone with density of 2.68 t/m^3, which was deemed the most suitable empirical formula for the present study.

The analysis was supported by video recordings and photographs such as those presented in Figure 10, covering the sections of interest. These were taken during the same tests discussed in the previous sub-section. The comparison was made by matching the results, for equivalent wave conditions, of the physical model with and without the WEC, as seen in Figure 11.

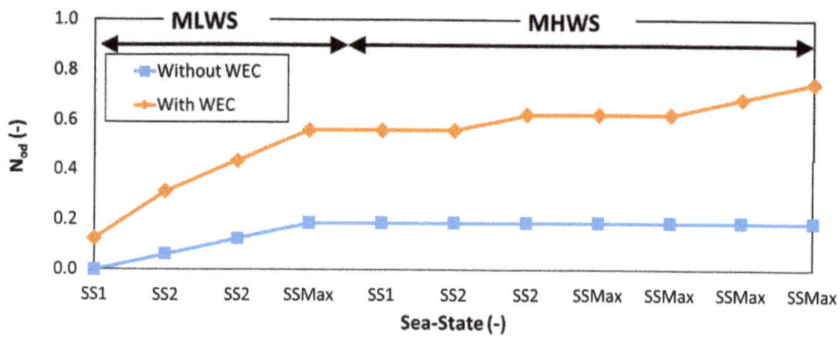

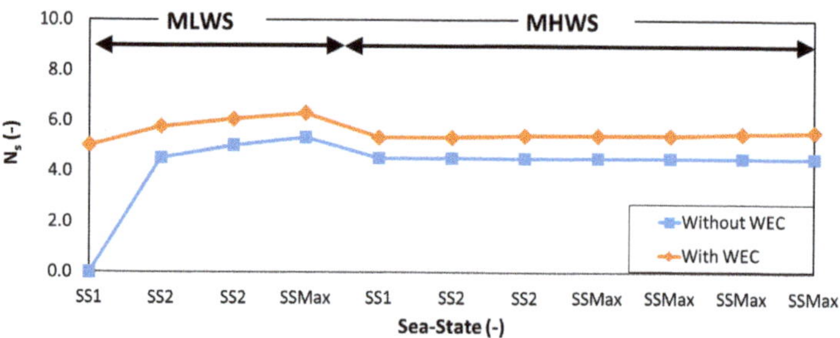

Figure 11. N_{od} and N_S values for the extreme sea-state tests, with and without the WEC, for the toe berm blocks.

Regarding the analysis of the toe berm blocks, the values are cumulative and for the whole toe berm, following the order of the tests from MLWS to MHWS (Table 1). Even so, it is noticeable that, without the WEC, there is a gradual increment of the N_{od} for MLWS level as the significant wave height increases, yet it remains unchanged for MHWS (0.19), as it reaches an equilibrium state, as discussed previously. In contrast, the N_{od} values with the WEC are much higher for both sea levels, with a maximum of 0.56 and 0.75 for MLWS and MHWS, respectively. Furthermore, the increment curve is less steep for MLWS, and block motion is recorded during MHWS tests with the WEC. Lastly, the stability number is slightly higher, overall, with the introduction of the WEC. The values range between 1.97 and 2.47/2.83 (MLWS/MHWS) and 0.00 and 2.09/2.30 (MLWS/MHWS) in the presence and absence of the WEC, respectively. Both limits are within the recommended threshold of 1 to 4 for statically stable breakwaters, a category into which rubble-mound breakwaters are included [20,21]. However, the toe berm may safely allow values above that range. Concerning the armour layer, no significant displacement of units was observed, leading to N_{od} values always equal to zero. As such, little to no damage is expected to occur at the armour layer under current design conditions, either with or without the WEC. Even so, the toe berm has not reached a complete stability state, and additional damage could occur if more tests were to be carried out. Consequently, this does not remove the possibility of structural failure, although no block motion was recorded for the armour layer blocks, with and without the WEC. The increment in the reflection coefficients also hints towards the potential generation of a stronger seaward current responsible for the additional damage at the toe berm, with the HWEC. This should be further assessed in follow-up studies.

For Antifer cubes on the toe berm, the reference value of damage start is 0.5, for N_{od}, and 2.42 to 2.67, for a corresponding N_s, considering the MLWS and the MHWS, respectively [21]. While the scenario without the WEC remains below these thresholds, even after the last test with MHWS, the integration of the WEC leads to an increase in N_{od} and N_S slightly beyond the upper limits, even when the whole cross section is considered. The damage number further increases and reaches a maximum of 0.75, an intermediate value between the damage start level and the acceptable damage level, which has a threshold of 2. The corresponding N_s is 2.83, for the obtained damage number, and 3.28, for the MHWS threshold (2.98 for MLWS). Even so, for the whole cross section, neither case leads to a situation where the N_{od} reaches or exceeds the "severe damage level", since the values are all far below 4 (3.31 to 3.64, for a corresponding N_s). In summary, the N_{od} and N_s in the absence of the WEC are, in accordance to Table VI-5-46 of the CEM [21], in the "no damage" case. With the WEC, "no damage" is mostly associated with the MLWS, while the MHWS yields only the "acceptable damage" case.

Nevertheless, it is important to take into account that the breakwater with the WEC integrated has a non-uniform cross-section, and the toe berm presents a distinct behaviour in front of the hybrid WEC (Figure 10). Figure 12 presents the evolution of the N_{od} parameter for the two stretches: in front of the WEC and remaining stretch, with maximum values of 0.87 and 0.62, respectively. It can be seen that while for the conventional cross-section N_{od} is safely below the start of damage threshold, at first, in front of the hybrid WEC, the toe berm is very close to the moderate damage condition. For the final layout, however, the damage number in front of the WEC stabilizes, but it increases to 0.62 for the adjacent cross-section, thus overcoming the moderate damage condition. These preliminary results reveal that methodologies that were developed for the analysis of conventional breakwater structures should be applied with care when WECs are integrated, forming non-uniform structures with completely different behaviours that should be analysed independently, to ensure the overall stability of the structure. The evaluation should also account for a significant duration and range of extreme wave conditions, given the evolution of N_{od} attributed to each reference stretch of the toe berm.

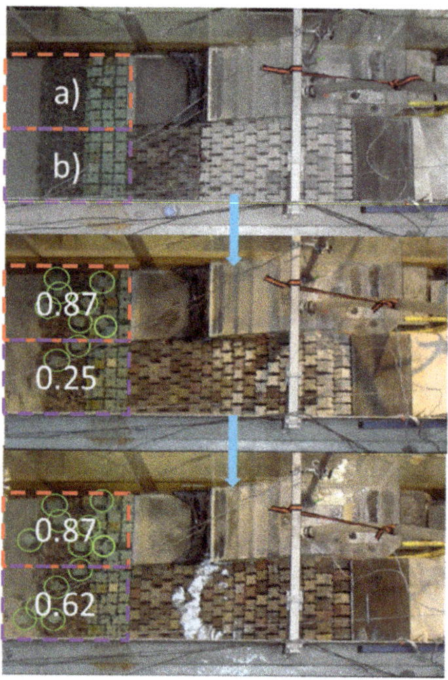

Figure 12. N_{od} values for the extreme sea-state tests with the WEC integrated in the breakwater, for the area: (**a**) in front of the WEC and (**b**) adjacent to the WEC, going from the initial MLWS to the final MHWS test.

4. Conclusions

In this paper, the integration of a hybrid OWEC–OWC wave energy converter device into the north breakwater at the Port of Leixões was analysed. The influence of such a combination in terms of structural stability and functionality of the rubble-mound breakwater was addressed and discussed, following the recommendations from the literature on the subject.

From a functional perspective, the integration of the hybrid concept denotes a beneficial contribution to overtopping volume mitigation, with an average reduction of about 50% for the tested conditions. Furthermore, the dimensionless discharge was compared to the overtopping prediction formula from EurOtop II. The results with and without the WEC both yielded roughness factors above the recommended value for Antifer blocks, given the regular placement of the armour layer blocks and the smooth surface of the WEC structure. In contrast, the overtopping results exceed the recommendations of the Rock Manual, the EurOtop II Manual, and the Coastal Engineering Manual regarding safety towards pedestrians (including trained staff), vehicles, equipment, and vessel access, being at least two orders of magnitude above the thresholds. Hence, if the function or use of the structure (i.e., the north breakwater extension) is changed in the future, it may be necessary to introduce particular modifications in the structural solution, namely, in terms of the crest's shape (assuming the implantation of the WEC) and armour layer block layout. This is dependent upon the intended usage of the inner area of the breakwater extension and the crest, which is currently foreseen to be very limited. Concerning wave reflections, the introduction of the HWEC led to an increase/decrease in K_r for MLWS/MHWS, respectively. These were a consequence of the HWEC's design, namely, the sloped reservoirs (MLWS) and the upper part of the HWEC module, which exhibit an essentially vertical configuration (MHWS). This should be carefully analysed in an eventual deployment stage, as it could affect the local navigation conditions and wave regime.

The structural stability analysis indicates that the introduction of the WEC has a negative impact on the stability of the toe berm of the breakwater, although the armour layer was not affected in a significant way (no relevant movements in the armour layer blocks). The level of damage is greater for the case with the WEC device integrated into the breakwater, yet the damage number and stability number values are lower than the failure level threshold when the whole toe berm is considered. At worst, an "acceptable damage" is obtained, and mostly for the MHWS with the WEC. A more specific analysis, dividing the toe berm in two parts, reveals higher damage numbers in front of the WEC, particularly for the MLWS, highlighting the fact that traditional damage assessment methods should be applied with care when non-conventional structures are analysed, such as breakwaters with integrated WEC. The original toe berm solution should also be redesigned, in order to ensure that the risk of failure is mitigated, as well as that of scouring. This should be a feasible option, as the inherent costs are expected to be low in comparison with the costs of the armour layer.

Author Contributions: Conceptualization, methodology, validation, formal analysis, investigation: D.C., T.C.-C., P.R.-S. and F.T.-P.; software: D.C., T.C.-C. and P.R.-S.; visualization, data curation: D.C. and T.C.-C.; writing—original draft preparation/review and editing: D.C.; writing—review and editing: T.C.-C., P.R.-S. and F.T.-P.; supervision, resources, project administration: P.R.-S. and F.T.-P. All authors have read and agreed to the published version of the manuscript.

Funding: This research was funded by the OCEANERA-NET project SE@PORTS—Sustainable Energy at SeaportS—with the references OCEANERA/0004/2016 and OCEANERA/0003/2016, under the frame of FCT, the Portuguese Foundation for Science and Technology, by the project PORTOS—Ports Towards Energy Self-Sufficiency—with the reference EAPA_784/2018 and co-financed by the Interreg Atlantic Area Program through the European Regional Development Fund and by the OCEANERA-NET COFUND project WEC4Ports—A Hybrid Wave Energy Converter for Ports—with the reference OCEANERA/0004/2019, under the frame of FCT.

Institutional Review Board Statement: Not applicable.

Informed Consent Statement: Not applicable.

Acknowledgments: The authors are in debt to LNEC, which lent the Antifer units used to materialize the armour layer of the reference breakwater model, to ELEVO for providing the rubble mound for the core and internal layers of that breakwater, and to the Port Authority of Douro, Leixões, and Viana do Castelo, for the information provided to characterize the case study. João Henriques (from Instituto Superior Técnico) provided important contributions to the testing and control of the PTO system used in the OWC module. The authors wish also to acknowledge the remaining partners of the SE@Ports and PORTOS consortium, namely, INEGI, who fabricated the physical model, and to the Port Authority of Douro, Leixões, and Viana do Castelo, for all the information provided to characterize the case study.

References

1. Hiranandani, V. Sustainable development in seaports: A multi-case study. *WMU J. Marit. Aff.* **2014**, *13*, 127–172. [CrossRef]
2. Murphy, E.; King, E.A. An assessment of residential exposure to environmental noise at a shipping port. *Environ. Int.* **2014**, *63*, 207–215. [CrossRef]
3. Pittaluga, I.; Borelli, D.; Repetto, S. Noise Pollution Management in Ports: A brief Review and the EU MESP Project Experience. In Proceedings of the 21st International Congress on Sound and Vibration, Beijing, China, 13–17 July 2014; p. 8.
4. Winnes, H.; Styhre, L.; Fridell, E. Reducing GHG emissions from ships in port areas. *Res. Transp. Bus. Manag.* **2015**, *17*, 73–82. [CrossRef]
5. Schipper, C.A.; Vreugdenhil, H.; de Jong, M.P.C. A sustainability assessment of ports and port-city plans: Comparing ambitions with achievements. *Transp. Res. Part D Transp. Environ.* **2017**, *57*, 84–111. [CrossRef]
6. Becker, A.; Inoue, S.; Fischer, M.; Schwegler, B. Climate change impacts on international seaports: Knowledge, perceptions, and planning efforts among port administrators. *Clim. Chang.* **2012**, *110*, 5–29. [CrossRef]

7.   Becker, A.H.; Acciaro, M.; Asariotis, R.; Cabrera, E.; Cretegny, L.; Crist, P.; Esteban, M.; Mather, A.; Messner, S.; Naruse, S.; et al. A note on climate change adaptation for seaports: A challenge for global ports, a challenge for global society. *Clim. Chang.* **2013**, *120*, 683–695. [CrossRef]

8.   Acciaro, M.; Vanelslander, T.; Sys, C.; Ferrari, C.; Roumboutsos, A.; Giuliano, G.; Lam, J.S.L.; Kapros, S. Environmental sustainability in seaports: A framework for successful innovation. *Marit. Policy Manag.* **2014**, *41*, 480–500. [CrossRef]

9.   Lam, J.S.L.; Notteboom, T. The Greening of Ports: A Comparison of Port Management Tools Used by Leading Ports in Asia and Europe. *Transp. Rev.* **2014**, *34*, 169–189. [CrossRef]

10.   Puig, M.; Wooldridge, C.; Darbra, R.M. Identification and selection of Environmental Performance Indicators for sustainable port development. *Mar. Pollut. Bull.* **2014**, *81*, 124–130. [CrossRef]

11.   Vicinanza, D.; Contestabile, P.; Quvang Harck Nørgaard, J.; Lykke Andersen, T. Innovative rubble mound breakwaters for overtopping wave energy conversion. *Coast. Eng.* **2014**, *88*, 154–170. [CrossRef]

12.   Vicinanza, D.; Lauro, E.D.; Contestabile, P.; Gisonni, C.; Lara, J.L.; Losada, I.J. Review of Innovative Harbor Breakwaters for Wave-Energy Conversion. *J. Waterw. Port Coast. Ocean Eng.* **2019**, *145*, 03119001. [CrossRef]

13.   Di Lauro, E.; Lara, J.L.; Maza, M.; Losada, I.J.; Contestabile, P.; Vicinanza, D. Stability analysis of a non-conventional breakwater for wave energy conversion. *Coast. Eng.* **2019**, *145*, 36–52. [CrossRef]

14.   Bruun, P. *Design and Construction of Mounds for Breakwaters and Coastal Protection*; Elsevier: Amsterdam, The Netherlands, 2013.

15.   Naty, S.; Viviano, A.; Foti, E. Wave Energy Exploitation System Integrated in the Coastal Structure of a Mediterranean Port. *Sustainability* **2016**, *8*, 1342. [CrossRef]

16.   Arinaga, R.A.; Cheung, K.F. Atlas of global wave energy from 10 years of reanalysis and hindcast data. *Renew. Energy* **2012**, *39*, 49–64. [CrossRef]

17.   Gunn, K.; Stock-Williams, C. Quantifying the global wave power resource. *Renew. Energy* **2012**, *44*, 296–304. [CrossRef]

18.   Van der Meer, J.W. Geometrical design of coastal structures. In *Seawalls Dikes Revetments*; Pilarczyk, K.W., Ed.; Balkema: Rotterdam, The Netherlands, 1998; p. 15.

19.   Gerding, E. Toe structure stability of rubble mound breakwaters. In *Delft and Delft Hydraulics Report H1874*; Delft University of Technology: Delft, The Netherlands, 1993.

20.   CIRIA; Engineering CCFC; CETMEF. *The Rock Manual: The Use of Rock in Hydraulic Engineering*; Construction Industry Research & Information Association: London, UK, 2007.

21.   USACE. Coastal Engineering Manual Part VI: Design of Coastal Project Elements. In *Place of Publication Not Identified*; Books Express Publishing: Berkshire, UK, 2012.

22.   Van Gent, M.R.A. Rock stability of rubble mound breakwaters with a berm. *Coast. Eng.* **2013**, *78*, 35–45. [CrossRef]

23.   Van Gent, M.R.A.; van der Werf, I.M. Rock toe stability of rubble mound breakwaters. *Coast. Eng.* **2014**, *83*, 166–176. [CrossRef]

24.   Etemad-Shahidi, A.; Bali, M. Stability of rubble-mound breakwater using H50 wave height parameter. *Coast. Eng.* **2012**, *59*, 38–45. [CrossRef]

25.   Van der Meer, J.W. Stability of breakwater armour layers—design formulae. *Coast. Eng.* **1987**, *11*, 219–239. [CrossRef]

26.   Hudson, R.Y. *Design of Quarry Stone Cover Layer for Rubble Mound Breakwaters*; Waterways Experiment Station, Coastal Engineering Research Centre: Vicksburg, MS, USA, 1958.

27.   Vidal, C.; Medina, R.; Lomónaco, P. Wave height parameter for damage description of rubble-mound breakwaters. *Coast. Eng.* **2006**, *53*, 711–722. [CrossRef]

28.   Kralli, V.-E.; Theodossiou, N.; Karambas, T. Optimal Design of Overtopping Breakwater for Energy Conversion (OBREC) Systems Using the Harmony Search Algorithm. *Front Energy Res.* **2019**, *7*, 11. [CrossRef]

29.   Bingham, H.B.; Ducasse, D.; Nielsen, K.; Read, R. Hydrodynamic analysis of oscillating water column wave energy devices. *J. Ocean. Eng. Mar. Energy* **2015**, *1*, 405–419. [CrossRef]

30.   Bruce, T.; van der Meer, J.W.; Franco, L.; Pearson, J.M. Overtopping performance of different armour units for rubble mound breakwaters. *Coast. Eng.* **2009**, *56*, 166–179. [CrossRef]

31.   Van der Meer, J.W.; Verhaeghe, H.; Steendam, G.J. The new wave overtopping database for coastal structures. *Coast. Eng.* **2009**, *56*, 108–120. [CrossRef]

32.   De Rouck, J.; Verhaeghe, H.; Geeraerts, J. Crest level assessment of coastal structures—General overview. *Coast. Eng.* **2009**, *56*, 99–107. [CrossRef]

33.   Allsop, W.; Bruce, T.; Pearson, J.; Besley, P. Wave overtopping at vertical and steep seawalls. *Proc. Inst. Civ. Eng.-Marit. Eng.* **2005**, *158*, 103–114. [CrossRef]

34.   Romano, A.; Bellotti, G.; Briganti, R.; Franco, L. Uncertainties in the physical modelling of the wave overtopping over a rubble mound breakwater: The role of the seeding number and of the test duration. *Coast. Eng.* **2015**, *103*, 15–21. [CrossRef]

35.   Losada, I.J.; Lara, J.L.; Guanche, R.; Gonzalez-Ondina, J.M. Numerical analysis of wave overtopping of rubble mound breakwaters. *Coast. Eng.* **2008**, *55*, 47–62. [CrossRef]

36.   Cavallaro, L.; Dentale, F.; Donnarumma, G.; Foti, E.; Musumeci, R.E.; Carratelli, E.P. Rubble Mound Breakwater Overtopping: Estimation of the Reliability of a Numerical 3D Simulation. *Coast. Eng. Proc.* **2012**, *1*, 8. [CrossRef]

37.   Briganti, R.; Bellotti, G.; Franco, L.; De Rouck, J.; Geeraerts, J. Field measurements of wave overtopping at the rubble mound breakwater of Rome–Ostia yacht harbour. *Coast. Eng.* **2005**, *52*, 1155–1174. [CrossRef]

38. Troch, P.; Geeraerts, J.; Van de Walle, B.; De Rouck, J.; van Damme, L.; Allsop, W.; Franco, L. Full-scale wave-overtopping measurements on the Zeebrugge rubble mound breakwater. *Coast. Eng.* **2004**, *51*, 609–628. [CrossRef]
39. Van der Meer, J.W.; Allsop, N.W.H.; Bruce, T.; De Rouck, J.; Kortenhaus, A.; Pullen, T.; Schüttrumpf, H.; Troch, P.; Zanuttigh, B. EurOtop: Manual on wave overtopping of sea defences and related structures. *EurOtop* **2018**. Available online: http://www.overtopping-manual.com/ (accessed on 1 August 2021).
40. Sigurdarson, S.; van der Meer, J.W. Wave Overtopping at Berm Breakwaters in Line with EurOtop. *Coast. Eng. Proc.* **2012**, *33*, 12. [CrossRef]
41. LNEC. *Estudos em Modelo Físico e Numérico do Prolongamento do Quebra-mar Exterior e das Acessibilidades Marítimas do Porto de Leixões*; Laboratório Nacional de Engenharia Civil (LNEC): Lisboa, Portugal, 2017.
42. Calheiros-Cabral, T.; Clemente, D.; Rosa-Santos, P.; Taveira-Pinto, F.; Ramos, V.; Morais, T.; Cestaro, H. Evaluation of the annual electricity production of a hybrid breakwater-integrated wave energy converter. *Energy* **2020**, *213*, 17. [CrossRef]
43. Cabral, T.; Clemente, D.; Rosa-Santos, P.; Taveira-Pinto, F.; Morais, T.; Belga, F.; Cestaro, H. Performance Assessment of a Hybrid Wave Energy Converter Integrated into a Harbor Breakwater. *Energies* **2020**, *13*, 236. [CrossRef]
44. Pecher, A.; Kofoed, J.P. (Eds.) *Handbook of Ocean Wave Energy*; Springer International Publishing: Cham, Switzerland, 2017; Volume 7. [CrossRef]
45. Iuppa, C.; Cavallaro, L.; Musumeci, R.E.; Vicinanza, D.; Foti, E. Empirical overtopping volume statistics at an OBREC. *Coast. Eng.* **2019**, *152*, 103524. [CrossRef]
46. Margheritini, L.; Vicinanza, D.; Kofoed, J.P. *Overtopping Performance of Sea Wave Slot Cone Generator. Coasts, Marine Structures and Breakwaters: Adapting to Change*; Thomas Telford Ltd.: London, UK, 2010; pp. 750–761. [CrossRef]
47. Contestabile, P.; Ferrante, V.; Di Lauro, E.; Vicinanza, D. Prototype Overtopping Breakwater for Wave Energy Conversion at Port of Naples. In Proceedings of the 26th International Ocean and Polar Engineering Conference, Rhodes, Greece, 26 June– 2 July 2016; International Society of Offshore and Polar Engineers: Rhodes, Greece; p. 6.
48. Torre-Enciso, Y.; Ortubia, I.; de Aguileta, L.I.L.; Marqués, J. Mutriku Wave Power Plant: From the Thinking out to the Reality. In Proceedings of the 8th European Wave and Tidal Energy Conference, Uppsala, Sweden, 7–10 September 2009; p. 11.
49. Arena, F.; Romolo, A.; Malara, G.; Fiamma, V.; Laface, V. *The First Full Operative U-OWC Plants in the Port of Civitavecchia*; Ocean Renew. Energy; ASME: Trondheim, Norway, 2017; p. V010T09A022. Volume 10. [CrossRef]
50. Koutrouveli, T.I.; Di Lauro, E.; das Neves, L.; Calheiros-Cabral, T.; Rosa-Santos, P.; Taveira-Pinto, F. Proof of Concept of a Breakwater-Integrated Hybrid Wave Energy Converter Using a Composite Modelling Approach. *J. Mar. Sci. Eng.* **2021**, *9*, 226. [CrossRef]
51. Cabral, T.; Clemente, D.; Rosa-Santos, P.; Taveira-Pinto, F.; Belga, F.; Morais, T. Preliminary Assessment of the Impact of a Hybrid Wave Energy Converter in the Stability and Functionality of a Rubble-Mound Breakwater. In Proceedings of the Coastal Structures Conference, Hannover, Germany, 29 September – 2 October 2019; pp. 1141–1151. [CrossRef]
52. Rosa-Santos, P.; Taveira-Pinto, F.; Clemente, D.; Cabral, T.; Fiorentin, F.; Belga, F.; Morais, T. Experimental Study of a Hybrid Wave Energy Converter Integrated in a Harbor Breakwater. *J. Mar. Sci. Eng.* **2019**, *7*, 33. [CrossRef]
53. Mansard, E.P.D.; Funke, E.R. The Measurement of Incident and Reflected Spectra Using a Least Squares Method. In Proceedings of the 17th International Conference on Coastal Engineering, Sydney, Australia, 23–28 March 1980. [CrossRef]

Journal of
*Marine Science
and Engineering*

Article

Theoretical Evaluation of the Hydrodynamic Characteristics of Arrays of Vertical Axisymmetric Floaters of Arbitrary Shape in front of a Vertical Breakwater

Dimitrios N. Konispoliatis [1,*], Spyridon A. Mavrakos [1,2] and Georgios M. Katsaounis [1]

[1] Laboratory for Floating Structures and Mooring Systems, Division of Marine Structures, School of Naval Architecture and Marine Engineering, National Technical University of Athens, 9 Heroon Polytechniou Avenue, GR 157-73 Athens, Greece; mavrakos@naval.ntua.gr (S.A.M.); katsage@mail.ntua.gr (G.M.K.)

[2] Hellenic Centre for Marine Research, Director and President, 190 13 Anavyssos, Greece

* Correspondence: dkonisp@naval.ntua.gr

Received: 12 December 2019; Accepted: 16 January 2020; Published: 20 January 2020

Abstract: The present paper deals with the analytical evaluation of the hydrodynamic characteristics of an array of vertical axisymmetric bodies of arbitrary shape, placed in front of a reflecting vertical breakwater, which can be conceived as floaters for wave power absorption. At the first part of the paper, the hydrodynamic interactions between the floaters and the adjacent breakwater are exactly taken into account using the method of images, whereas, the interaction phenomena between the floaters of the array are estimated using the multiple scattering approach. For the solution of the problem, the flow field around each floater of the array is subdivided into ring-shaped fluid regions, in each of which axisymmetric eigenfunction expansions for the velocity potential are made. In the second part of the paper, extensive theoretical results are presented concerning the exciting wave forces and the hydrodynamic coefficients for various arrays' arrangements of axisymmetric floaters. The aim of the study is to show parametrically the effect that the vertical breakwater has on the hydrodynamic characteristics of each particular floater.

Keywords: vertical axisymmetric floaters; arbitrary shape; breakwater; diffraction and radiation problem; hydrodynamic characteristics; added mass; damping coefficient

---

1. Introduction

Within the context of the linearized theory of water waves, a variety of methods have been devised for the calculation of hydrodynamic interaction phenomena within arrays of floating axisymmetric bodies having vertical symmetry axis. These methods have found application in many areas including the dynamics of oil & gas offshore platforms, the design of floating airports and other maritime structures. The application of most concern in the present paper is the use of arrays of vertical axisymmetric floaters in front of a vertical breakwater for absorbing wave power.

Several hundreds of patents related to harvesting of wave energy have been in existence by the late 20th century [1–4]. However, despite the increase of interest and awareness for wave energy absorption, the development from concept to commercial stage has been found to be a difficult, slow and expensive process. The main obstacle in harvesting the wave power is the high energy cost, related mainly to the survivability of the wave energy converter (WEC) and its critical components and sub-components (i.e., power take off system, mooring system, power electronics gearbox, etc.,) at the demanding offshore environmental conditions (i.e., extreme weather conditions, salt environment, etc.) [5]. Another aspect for the high energy cost is the often lack of development of an offshore grid infrastructure to transport the electricity from renewable offshore energy sources to centers of consumption and storage [6]. Furthermore, the uncertainties in identifying and mitigating the environmental impact of the WEC's life-cycle operation along with the lack of current licensing and consenting procedure lead the developers to face stringent and costly monitoring requirements before and after consent, increasing the wave energy cost [7].

Aiming at overcoming the aforementioned bottlenecks several parameters related to the WEC characteristics have been up to date examined. Representative examples are: (a) the WEC geometrical characteristics optimization, in the scope of harnessing maximum wave energy at the installation location; (b) the optimization of the WEC's characteristics with respect to their mechanical components, to withstand the demanding environmental conditions as well as to reduce the energy losses associated with the transformation of the wave power into electricity; and (c) the installation of WECs close to other near- or on-shore maritime structures such as a breakwater; a harbor or a pier, so as to use the already developed electric grid, reducing in parallel the environmental impact of the WEC's operation [8].

Looking towards the possible advantages provided by installing of WEC devices in near- or on-shore areas or close to other maritime structures, several studies have been presented in the literature. Indicatively, in [9] the most representative existent wave energy converters were evaluated in various offshore and near-shore areas, whereas, in [10] the installation of a WEC device in a port with dual operation, i.e., both as a breakwater as well a wave energy device was examined.

Furthermore, the effect of a reflecting vertical wall on the WECs' behavior has been investigated in several studies in the last years. The majority of them are dealing with WECs installed either at a certain distance from the wall or integrated at it. More specifically, in [11] the performance characteristics of an array of five wave energy heaving converters placed in front of a reflecting vertical breakwater have been numerically studied whereas, in [12] the performance of an array of heaving WECs, coupled with DC generators, in front of a breakwater, were numerically and experimentally investigated. Furthermore, solution methods concerning the wave diffraction and radiation problems for the case of a truncated cylinder, in front of a vertical wall have been presented in the literature [13–16], whereas, in [17] the efficiency of a heaving point absorber in front of a vertical wall in regular and irregular seas was studied, based on different floater geometries and wave heading angles. In [18] the possibility of using oscillating water column (OWC) converters for reducing the wave reflection from vertical breakwaters was explored.

As far as investigations concerning the behavior of WECs integrated at a breakwater, theoretical hydrodynamic studies of an oscillating water column device (OWC) placed at the tip of a breakwater and along a straight coast were developed in [19,20]. Moreover, numerical analysis and experimental investigation into OWCs integrated at a flat breakwater have been presented in [21] whereas, in [22,23] a feasibility study of an OWC device integrated into a port in the Mediterranean Sea was presented. In [24,25] multiple OWCs structures integrated into floating breakwaters were investigated. Recently, in [26] an analytical study of a pile supported OWC breakwater was presented, whereas, in [27,28] the performance of a WEC integrated into a breakwater was investigated, analytically and experimentally.

Despite the difficulties associated with the placement of WECs in front of a breakwater in terms of installation, mooring and maintenance issues, such converters appear to outperform the integrated into the breakwater counterparts. Their main advantage is the possibility they offer to have them installed at different distances from the vertical wall in dependence from the prevailing sea conditions at the particular installation site, thus optimizing their performance taking advantage of the particular, site-dependent interaction phenomena between the floaters and the breakwater. In addition, WECs placed in front of a breakwater can act as a protecting mole to the incoming waves, reducing the intensity of wave action on the shore at severe environment conditions [8].

As a common wave energy device, an oscillating buoy has been proved to be an effective WEC for wave energy extraction due to its favorable properties: ease of installation, economic operation and manufacturing processes [29–32]. The converter is composed by a base moored to the sea bottom and an arbitrary shape floater. The relative motion of the buoyant top to the base is converted into electrical power by the power-take-off (PTO) mechanism. The PTO system can be a high-pressure oil system, converting the floater's motions into hydraulic energy and then converted into electricity through hydraulic motors, or a linear electrical generator [33,34].

The main objective of the present paper is to evaluate the hydrodynamic characteristics (exciting wave forces, hydrodynamic coefficients) of an array of vertical axisymmetric floaters of arbitrary shape that are floating in finite depth waters in front of a vertical and fully reflecting breakwater of infinite length. Towards this goal, the breakwater's effect on the floaters' hydrodynamic coefficients (i.e., hydrodynamic added mass and damping coefficients) and on their exciting wave loads is derived by accounting for the hydrodynamic interaction phenomena both between the bodies of the array as well as between them and the adjacent breakwater.

As far as the simulation of the vertical breakwater is concerned, the method of images is used to describe the fluid flow around the array in front of it. According to this method, the problem of N number of floaters in front of the vertical breakwater is equivalent to the one of an array of $2N$ number of floaters consisting of the initial ones and their image virtual devices with respect to the breakwater that are exposed to the action of surface waves without, however, the presence of the breakwater. The method of images has been initially applied to tackle the diffraction and radiation problems of single or array of cylinders in channels, simulating accurately the reflections from the side walls [35–38]. Later, this method has been also used to simulate the effect of a vertical wall on an array of floating bodies placed in front of it [11,15,16].

Furthermore, considering the solution of the relevant linearized diffraction and radiation problems, the hydrodynamic interference effects between the floaters in the array are evaluated within the context of potential flow theory using single-body hydrodynamic characteristics of the individual bodies and the method of multiple scattering [39–42]. In this formulation, the incident wave potential and various orders of propagating and evanescent wave modes radiated and scattered from all the bodies in the array are superposed to obtain exact series representations of the total wave field around each body of the configuration. As the boundary conditions are satisfied successively on each body in the array, there is no need to retain simultaneously the unknown partial wave amplitudes around all the floaters. As a result, a considerable reduction of the storage requirements in computer applications can be obtained, without, however, compromising the accuracy of the outcomes of the multiple scattering approach compared to other computational methods [43,44].

Finally, in the context of the present contribution, extensive theoretical results concerning the hydrodynamic characteristics (exciting wave forces, hydrodynamic mass and damping coefficients) for several shapes of the individual floaters and array configurations, are given in form of figures and in tabular form. The presented results show that the hydrodynamic characteristics of an array of vertical axisymmetric floaters in front of a vertical wall are evidently different from those in unbounded waters (i.e., without the presence of the breakwater) and their values are dependent form the wave number (i.e., wave frequency), floaters' geometry and arrays' configuration.

2. Hydrodynamic Formulation

An array of N vertical axisymmetric floaters, of arbitrary shape, placed in front of a breakwater at constant water depth d is considered. The floaters are exposed to the action of a plane incident wave train of frequency ω and amplitude A propagating at an angle θ with respect to the positive x-axis. A global, right-handed Cartesian co-ordinate system O-xyz is introduced with origin O located at the still water plane on the breakwater with its vertical axis Oz directed upwards. Moreover, N local cylindrical co-ordinate systems (r_q, θ_q, z_q), $q = 1, 2, \ldots, N$, are defined with origins at the intersection (X_q, Y_q) of the calm water surface with the vertical axis of symmetry of each body. Three different types of floating bodies are examined, (a) a conical floater; (b) a vertical cylindrical floater; and (c) a semi-spherical floater, as seen in Figure 1. All the examined floaters have an outer diameter of D and distance from the sea bed h, whereas the draught of the cylindrical part of the floater, in cases a and c, is denoted by h_1. The distance between the center of the closest to the wall floater and the breakwater is denoted by l_W (see Figure 1).

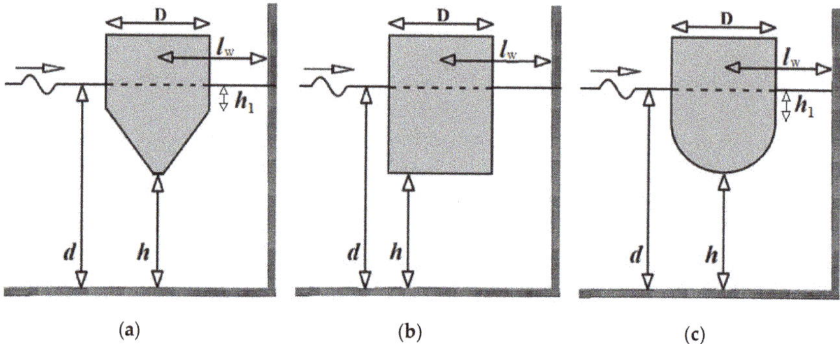

Figure 1. 2-D representations of the examined types of floaters in front of a breakwater: (**a**) a conical floater; (**b**) a vertical cylindrical floater; (**c**) a semi-spherical floater.

In order to describe the fluid flow around the N floaters of the array in front of the breakwater, the method of images is applied. According to this method, the problem under investigation can be traced back to an equivalent problem of bi-directional incident waves, one propagating at angle θ and one at angle $180-\theta$ incident on an array of $2N$ floaters, without the presence of the breakwater (see Figure 2).

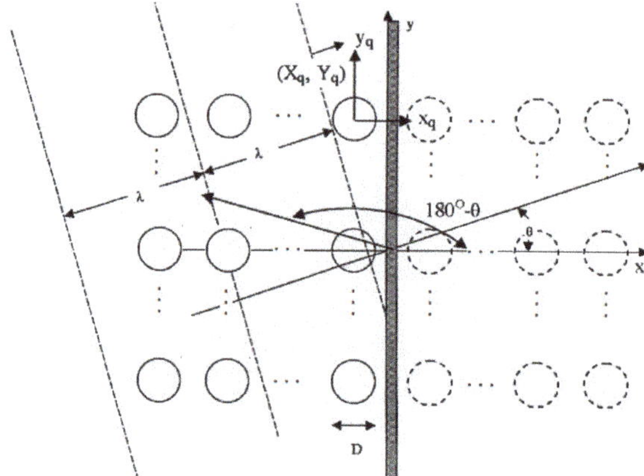

Figure 2. Plan view of the array of N floaters in front of the breakwater (image cylinders are denoted dashed).

Assuming that the flow is irrotational and inviscid and that the waves are of small slope, classical linearized water wave theory can be employed. The fluid flow around each floater q, $q = 1$, $\ldots$, $2N$, (including the initial and the image floaters) can be described by the potential function, $\Phi\left(r_q, \theta_q, z_q; t\right) = Re\left[\varphi^q\left(r_q, \theta_q, z_q\right)\right]$, and expressed, on the basis of linear modeling, as a superposition of incident φ_0, scattered, φ_s^q and radiated wave fields, φ_j^q, i.e.,

$$\varphi^q\left(r_q, \theta_q, z_q\right) = \varphi_0\left(r_q, \theta_q, z_q\right) + \varphi_s^q\left(r_q, \theta_q, z_q\right) + \sum_{q=1}^{2N}\sum_{j=1}^{5}\varphi_j^q\left(r_q, \theta_q, z_q\right) \tag{1}$$

Here, φ_j^q denotes the potential of the wave field induced by the forced oscillation of q floater in the j-th direction, the remaining ones considered restrained.

The velocity potential of the undisturbed incident wave propagating at an angle θ with respect to the positive x-axis, expressed in the co-ordinate frame of the q floater is:

$$\varphi_0\left(r_q, \theta_q, z_q\right) = -i\omega A \sum_{m=-\infty}^{\infty} i^m \Psi_{0,m}\left(r_q, z_q\right) e^{im\theta_q} \tag{2}$$

where:

$$\frac{1}{d}\Psi_{0,m}\left(r_q, z_q\right) = e^{ikl_{0q}\cos\left(\theta_{0q}-\theta\right)} \frac{Z_0(z)}{dZ_0'(0)} J_m\left(kr_q\right) e^{-im\theta} \tag{3}$$

Here, J_m is the m-th order Bessel function of the first kind; $\left(l_{0q}, \theta_{0q}\right)$ are the polar coordinates of the q floater center relative to the origin O of the global co-ordinate systems O–xyz and $Z_0(z)$ are orthonormal functions in $[0, d]$ defined as follows:

$$Z_0(z) = \left[\frac{1}{2}\left[1 + \frac{\sin h(2kd)}{2kd}\right]\right]^{-1/2} \cos h(k(z+d)) \tag{4}$$

In accordance to Equation (2) the diffraction, $\varphi_D^q = \varphi_0 + \varphi_s^q$, and radiation, φ_j^q, velocity potentials, around the q floater, when it is considered isolated, can be expressed in the co-ordinate system of body q as follows:

$$\varphi_D^q = -i\omega A \sum_{m=-\infty}^{\infty} i^m \Psi_{D,m}^q\left(r_q, z_q\right) e^{im\theta_q} \tag{5}$$

$$\varphi_j^q = \dot{x}_{j0}^q \sum_{m=-\infty}^{\infty} \Psi_{j,m}^q\left(r_q, z_q\right) e^{im\theta_q} \quad j = 1, 2, \ldots, 5 \tag{6}$$

Here, $\dot{x}_{j0}^q$ is the complex velocity amplitude of q floater's motion in the j-th direction.

The velocity potentials, $\varphi_k^q \ k = 1, \ldots, 5, D$, have to satisfy the Laplace equation within the entire fluid domain; the linearized boundary conditions at the free surface; the zero normal velocity on the sea bed; the kinematic conditions on the mean floater's wetted surface and an appropriate radiation condition at infinity stating that the disturbance propagation must be outgoing [41,42].

The unknown functions $\Psi_{D,m}$, $\Psi_{j,m}$ involved in Equations (5) and (6) can be established through the method of matched axisymmetric eigenfunctions expansions. According to this method, the flow field around the floater q is subdivided in coaxial ring-shaped fluid regions, denoted by I and $III_p, p = 1, \ldots, L$ (see Figure 3), in which different series expansions of the velocity potential can be established. These series representations are solutions of the Laplace equation and satisfy the kinematic boundary condition at the walls of the floater; the linearized condition at the free surface; the kinematic condition at the sea bed; and the radiation condition at infinity. Moreover, the velocity potentials and their derivatives must be continuous at the vertical boundaries of neighboring fluid regions [45].

By the way of example, the appropriate expansions for the velocity potential, $\Psi_{D,m}$, $\Psi_{j,m}$, in form of Fourier—Bessel series in the fluid domain I and $III_p, p = 1, \ldots, L$ are presented below.

(a)   Infinite ring element $I \ (r \geq \frac{D}{2}, \ 0 \leq z \leq d)$

$$\frac{1}{\delta_k}\Psi_{k,m}^q\left(r_q, z\right) = g_{k,m}^q\left(r_q, z\right) + \sum_{n=0}^{\infty} F_{k,mn}^q \frac{K_m\left(a_n r_q\right)}{K_m\left(\frac{a_n D}{2}\right)} Z_n(z) \tag{7}$$

for $k = D, 1, \ldots, 5$;

where:

$$g_{D,m}^{q}(r_q, z) = \left\{ J_m(kr_q) - \frac{J_m\!\left(\frac{kD}{2}\right)}{H_m\!\left(\frac{kD}{2}\right)} H_m(kr_q) \right\} \frac{Z_0(z)}{d\acute{Z}_0(d)} \tag{8}$$

and $\delta_D = \delta_1 = \delta_2 = \delta_3 = d$, $\delta_4 = \delta_5 = d^2$; H_m, K_m are the m-th order Hankel function of first kind and the modified Bessel function of second kind, respectively; $F_{k,mn}^{q}$ are the unknown Fourier coefficients to be determined by the solution procedure. Furthermore,

$$Z_n(z) = \left[\frac{1}{2}\left[1 + \frac{\sin(2a_n d)}{2a_n d} \right] \right]^{-1/2} \cos(a_n(z+d)), \; n \geq 1 \tag{9}$$

The eigenvalues a_n are roots of the transcendental equation: $\omega^2 + ga_n \tan(a_n d) = 0$, which possesses one imaginary, $a_0 = -ik$, $k > 0$ and infinite number of real roots.

(b) p-th ring element III_p of the q floater ($\alpha_p \leq r_p \leq \alpha_{p+1}$, $0 \leq z \leq h_p$, $p = 1, 2, \ldots, L$)

$$\frac{1}{\delta_k}\Psi_{k,m}^{q}(r_p, z) = g_{k,m}^{q}(r_p, z) + \sum_{n_p=0}^{\infty} \epsilon_{n_p} \left[R_{mn_p}(r_p)F_{k,mn_p}^{q} + R_{mn_p}^{*}(r_p)F_{k,mn_p}^{*q} \right] \cos\left(\frac{n_p \pi (z+d)}{h_p} \right) \tag{10}$$

Here: $g_{D,m}^{q}(r_p, z) = g_{1,m}^{q}(r_p, z) = g_{2,m}^{q}(r_p, z) = 0$; $g_{3,m}^{q}(r_p, z) = \frac{(z+d)^2 - (\frac{1}{2})r_p^2}{2h_p d}$; $g_{4,m}^{q}(r_p, z) = -g_{5,m}^{q}(r_p, z) = \frac{-r_p\left[(z+d)^2 - (\frac{1}{4})r_p^2\right]}{2h_p d^2}$; δ_k has been defined above; F_{k,mn_p}^{q}, F_{k,mn_p}^{*q} are Fourier coefficients to be determined by the solution procedure; ϵ_{n_p} is Neumann's symbol defined as: $\epsilon_{n_p} = 1$, for $n_p = 0$; otherwise $\epsilon_{n_p} = 2$.

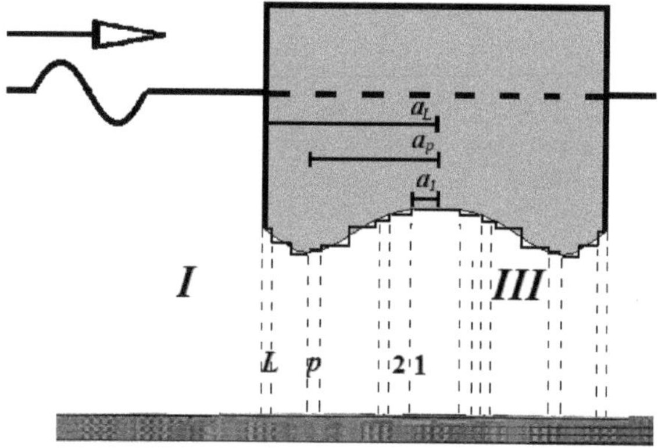

Figure 3. Discretization of the flow field around an axisymmetric arbitrary shape floater.

The R_{mn_p}, $R_{mn_p}^{*}$ terms express the radial dependence of the fluid's velocity in the p-th ring element. They can be written as:

$$R_{m0}(r_p) = \frac{\left(\frac{r_p}{\alpha_p}\right)^m - \left(\frac{\alpha_p}{r_p}\right)^m}{\left(\frac{\alpha_{p+1}}{\alpha_p}\right)^m - \left(\frac{\alpha_p}{\alpha_{p+1}}\right)^m}, \; R_{m0}^{*}(r_q) = \frac{\left(\frac{\alpha_{p+1}}{r_p}\right)^m - \left(\frac{r_p}{\alpha_{p+1}}\right)^m}{\left(\frac{\alpha_{p+1}}{\alpha_p}\right)^m - \left(\frac{\alpha_p}{\alpha_{p+1}}\right)^m}, \; n_p = 0 \tag{11}$$

$$R_{m0}(r_p) = \frac{K_m\!\left(\frac{n_p \pi \alpha_p}{h_p}\right)I_m\!\left(\frac{n_p \pi r_p}{h_p}\right) - I_m\!\left(\frac{n_p \pi \alpha_p}{h_p}\right)K_m\!\left(\frac{n_p \pi r_p}{h_p}\right)}{K_m\!\left(\frac{n_p \pi \alpha_p}{h_p}\right)I_m\!\left(\frac{n_p \pi \alpha_{p+1}}{h_p}\right) - I_m\!\left(\frac{n_p \eth \alpha_p}{h_p}\right)K_m\!\left(\frac{n_p \pi \alpha_{p+1}}{h_p}\right)}, \; n_p \neq 0 \tag{12}$$

$$R^*_{m0}(r_p) = \frac{I_m\left(\frac{n_p\pi\alpha_{p+1}}{h_p}\right)K_m\left(\frac{n_p\pi r_p}{h_p}\right) - K_m\left(\frac{n_p\pi\alpha_{p+1}}{h_p}\right)I_m\left(\frac{n_p\pi r_p}{h_p}\right)}{K_m\left(\frac{n_p\pi\alpha_p}{h_p}\right)I_m\left(\frac{n_p\pi\alpha_{p+1}}{h_p}\right) - I_m\left(\frac{n_p\pi\alpha_p}{h_p}\right)K_m\left(\frac{n_p\pi\alpha_{p+1}}{h_p}\right)}, n_p \neq 0 \tag{13}$$

Especially, for $p = 1$ (the middle fluid region underneath the q floater) the velocity potential is given as:

$$\frac{1}{\delta_k}\Psi^q_{k,m}(r_p,z) = g^q_{k,m}(r_p,z) + \sum_{n_1=0}^{\infty} \epsilon_{n_1} F^q_{k,mn_1} \frac{I_m\left(\frac{n_1 r_1}{h}\right)}{I_m\left(\frac{n_1\pi a_1}{h}\right)}\cos\left(\frac{n_1\pi(z+d)}{h}\right) \tag{14}$$

In accordance to the Equations (5) and (6) the diffraction and radiation velocity potential induced around any floater of the array can be written as:

$$\varphi^{qq}_D = -i\omega A \sum_{m=-\infty}^{\infty} i^m \Psi^{qq}_{D,m}(r_q,z_q)e^{im\theta_q} \tag{15}$$

$$\varphi^{qp}_j = \dot{x}^p_{j0} \sum_{m=-\infty}^{\infty} \Psi^{qp}_{i,m}(r_q,z_q)e^{im\theta_q} \tag{16}$$

Here $\Psi^{qq}_{D,m}$, is the diffraction potential due to the interference of the q floater with the incoming incident wave field and the scattered waves by all floaters in the array, whereas $\Psi^{qp}_{i,m}$ denote the radiation potentials around the q floater of the array due to the forced oscillation of the p floater.

In order to express the potentials in form of Equations (15) and (16) the multiple scattering method is applied, taking into consideration the interaction phenomena between the bodies of the array. The method which is applicable to arrays consisting of an arbitrary number of vertical axisymmetric bodies, having any geometrical arrangement and individual body geometry, has been described exhaustively in previous publications [41,42], thus it is not further elaborated here.

3. Hydrodynamic Reaction Forces

Having determined the diffraction and radiation velocity potentials around each floater of the array the exciting wave forces and the hydrodynamic reaction forces acting on the q floater; ($q = 1, 2, 2N$) can be obtained by:

$$F^q_{D,i} = -\iint_{S_q} i\omega\rho\varphi^{qq}_D n_i dS \tag{17}$$

$$F^{qp}_{ij} = -\iint_{S_q} i\omega\rho\varphi^{qp}_j n_i dS \tag{18}$$

Here, $F^q_{D,i}$ denotes the exciting wave force acting on the q floater in the i-th direction; F^{qp}_{ij} are the hydrodynamic reaction forces acting on the q floater in the i-th direction, due to the forced oscillation of the p floater in the j-th direction; ρ is the water density; S_q is the mean wetted surface of the q floater; n_i are the generalized normal components defined by: $n = (n_1, n_2, n_3); r \times n = (n_4, n_5); r$ being the position vector of a point on the wetted surface S_q with respect to the reference co-ordinate system of body q.

The hydrodynamic reaction forces F^{qp}_{ij} can be also written as [46]:

$$F^{qp}_{ij} = i\omega\left(a^{q,p}_{i,j} + \frac{i}{\omega}\beta^{q,p}_{i,j}\right)\dot{x}^p_{j0} \tag{19}$$

where $a^{q,p}_{i,j}$, $\beta^{q,p}_{i,j}$ are the added mass and damping coefficients, respectively, of the q floater in i-th direction due to the forced oscillation of the p floater in the j-th direction.

Based on the method of images the exciting forces acting on the q floater of an array of N floaters in front of a vertical breakwater exposed to the action of waves propagating at an angle θ, equal to the sum of the exciting forces acting on the initial q floater, for wave angles θ and 180–θ, assuming the presence of image floaters, with respect to the breakwater (i.e., total number of floaters 2N), without the presence of the vertical wall. Furthermore, the hydrodynamic coefficients $a_{i,j}^{q,p}$, $\beta_{i,j}^{q,p}$ of the q floater in i-th direction due the forced oscillation of the p floater in j-th mode of motion in front of the vertical breakwater can be derived, by summing up properly the motion-dependent hydrodynamic coefficients $a_{i,j}^{q,p}, b_{i,j}^{q,p}$ of the initial q floater in the i-th direction (i.e., $I = 1, \ldots, 5$) due to the forced oscillation of the p floater in the j-th direction (i.e., $j = 1, \ldots, 5$) with the corresponding hydrodynamic coefficients $a_{i,j}^{q,p'}, b_{i,j}^{q,p'}$ of the initial q floater due to the forced oscillation in the j-th direction of the image floater of the p body, denoted as p'. Table 1 denotes, indicatively, the determination of the added mass coefficient of the q floater of an array of N floaters in front of a vertical wall, using the image theory. The same formula is applied to the damping coefficients.

Table 1. Added mass coefficients of the q floater, $q = 1, \ldots, N$, in front of a vertical wall, using the image theory (the image floater of the p floater is denoted as p').

$a_{1,1}^{q,p} - a_{1,1}^{q,p'}$	$a_{1,2}^{q,p} + a_{1,2}^{q,p'}$	$a_{1,3}^{q,p} + a_{1,3}^{q,p'}$	$a_{1,4}^{q,p} + a_{1,4}^{q,p'}$	$a_{1,5}^{q,p} - a_{1,5}^{q,p'}$
$a_{2,1}^{q,p} - a_{2,1}^{q,p'}$	$a_{2,2}^{q,p} + a_{2,2}^{q,p'}$	$a_{2,3}^{q,p} + a_{2,3}^{q,pp'}$	$a_{2,4}^{q,p} + a_{2,4}^{q,p'}$	$a_{2,5}^{q,p} - a_{2,5}^{q,p'}$
$a_{3,1}^{q,p} - a_{3,1}^{q,p'}$	$a_{3,2}^{q,p} + a_{3,2}^{q,p'}$	$a_{3,3}^{q,p} + a_{3,3}^{q,p'}$	$a_{3,4}^{q,p} + a_{3,4}^{q,p'}$	$a_{3,5}^{q,p} - a_{3,5}^{q,p'}$
$a_{4,1}^{q,p} - a_{4,1}^{q,p'}$	$a_{4,2}^{q,p} + a_{4,2}^{q,p'}$	$a_{4,3}^{q,p} + a_{4,3}^{q,p'}$	$a_{4,4}^{q,p} + a_{4,4}^{q,p'}$	$a_{4,5}^{q,p} - a_{4,5}^{q,p'}$
$a_{5,1}^{q,p} - a_{5,1}^{q,p'}$	$a_{5,2}^{q,p} + a_{5,2}^{q,p'}$	$a_{5,3}^{q,p} + a_{5,3}^{q,p'}$	$a_{5,4}^{q,p} + a_{5,4}^{q,p'}$	$a_{5,5}^{q,p} - a_{5,5}^{q,p'}$

It should be also noted that the Table 1 fulfils the symmetry requirements of the added mass coefficients, i.e., $a_{1,3}^{q,p} + a_{1,3}^{q,p'} = a_{3,1}^{q,p} - a_{3,1}^{q,p'}$; $a_{1,5}^{q,p} - a_{1,5}^{q,p'} = a_{5,1}^{q,p} - a_{5,1}^{q,p'}$; $a_{4,2}^{q,p} + a_{4,2}^{q,p'} = a_{2,4}^{q,p} + a_{2,4}^{q,p'}$; etc.

4. Numerical Results

Initially, the theoretical results derived from the aforementioned analysis are compared with the available ones from the literature. The present analytical model is applied to the case of a single floating cylinder located in front of a vertical wall at finite water depth in order to compare the results with the ones of [15,16]. The examined cylinder of radius D/2 and draught 2($d - h$)/D = 0.5 is subjected to incident wave with an angle of attack $\theta = 0$ (i.e., the wave is propagating along the x axis), at a water depth 2d/D = 1.0, for various examined distances between the center of the cylinder and the vertical wall 2l_w/D (see Figure 1). The comparison is made, indicatively, in terms of the dimensionless surge

exciting forces (see Equation (17)), i.e., $F = \dfrac{\left| F_{D,1}^1 \right|}{\frac{\pi \rho g D^2 A}{4}}$; and the dimensionless hydrodynamic coefficients

(see Equation (19)), i.e., $\mu_i = \dfrac{a_{1,i}^{1,1}}{\frac{\pi \rho D^2 (d-h)}{4}}$, $i = 1, 3, 5$; $c_i = \dfrac{b_{1,i}^{1,1}}{\frac{\omega \pi \rho D^2 (d-h)}{4}}$, $i = 1, 3, 5$.

Figure 4 depicts the surge exciting forces for various examined distances between the device and the wall. An excellent correlation between the analytical results of the presented theoretical method and the analytical results from [15] can be obtained. In the Figure 5 the dimensionless surge hydrodynamic coefficients (i.e., hydrodynamic added mass and damping) of the floating cylinder due to its forced oscillations in surge, heave and pitch directions, are presented and compared, with also an excellent agreement, with the analytical results from [16].

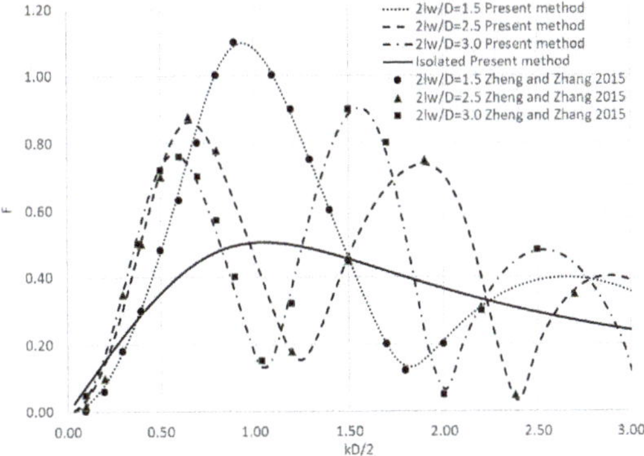

Figure 4. Dimensionless surge wave exciting forces against *k*D/2 for three distances between the floater and the vertical wall and for an isolated floater (i.e., without the presence of the breakwater). The results are compared with the analytical results of [15].

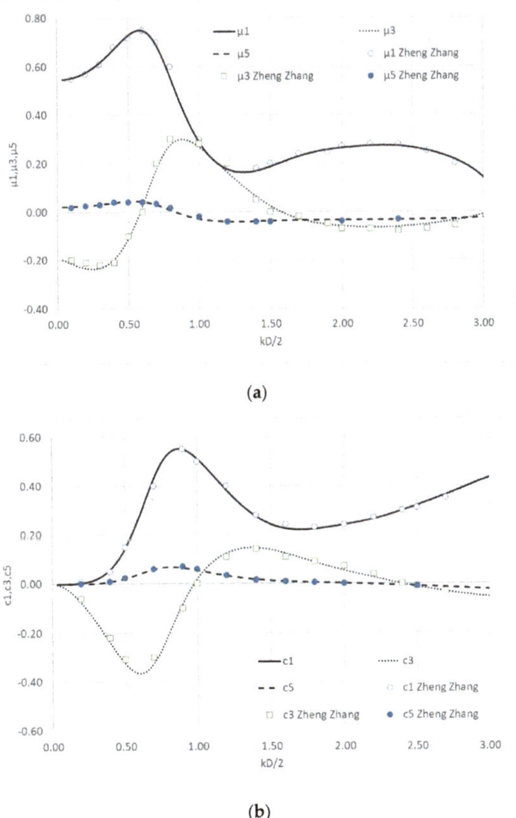

(a)

(b)

Figure 5. Dimensionless surge added mass (**a**) and surge damping coefficients (**b**) against *k*D/2 due to motions of the cylindrical body in surge, heave and pitch, for $l_w/D = 1$. The results are compared with the analytical ones of [16].

Next, three different array types of axisymmetric floaters with vertical symmetry axes are examined, (a) an array of five same conical floaters; (b) an array of five same vertical cylindrical floaters; and (c) an array of five same semi-spherical floaters (see Figure 1). The floaters are placed in front of a vertical breakwater of infinite length in three different array configurations, i.e., the devices are placed: (a) in a parallel direction to the wall; (b) in a rectangular arrangement in front of the wall; and (c) in a perpendicular direction to the wall; see Figure 6, C_1, C_2, C_3, respectively. The wave is assumed to propagate along the x-axis. The distance between the center of the closest to the wall floater and the breakwater is l_W, whereas the distance between adjacent bodies is l_b (in C_2 array—see Figure 6—the distance of the 4th from the 5th device and of the 1st from the 2nd device is $1.732l_b$). The examined floaters have a radius: $D/2$; draught: $(d - h)/D = 0.5$; and $D/(2h_1) = 10$; at a water depth: $d/D = 1.0$ (see Figure 1). The distances between the center of the cylinder and the vertical wall and between the adjacent cylinders are $l_w/D = 2$ and $l_b/D = 4$, respectively.

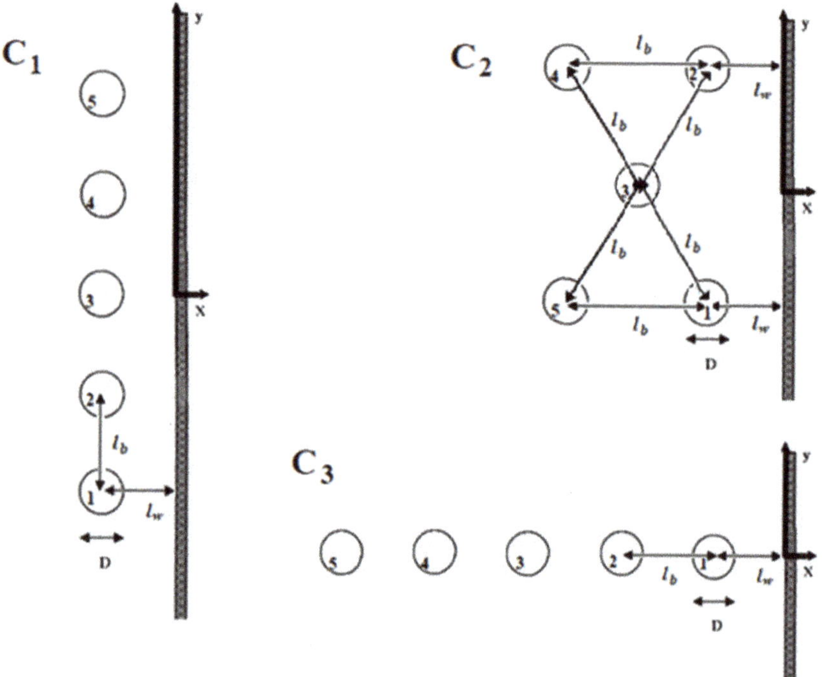

Figure 6. Three different examined array configurations concerning the position of each floater to the incoming wave and the vertical breakwater. The array C_1 is placed in a parallel direction to the wall; the array C_2 in a rectangular arrangement in front of the wall; and the array C_3 is placed in a perpendicular direction to the wall.

In the Figures 7 and 8 the dimensionless wave exciting forces in surge (see Equation (17)), i.e.,

$$F_x = \frac{\left|F_{D,1}^k\right|}{\frac{\pi \rho g D^2 A}{4}};$$

acting on the k floater, i.e., $k = 1,3$ of each examined array configuration in front of a breakwater (see Figure 6), are plotted against the corresponding values acting on the same floater of the array without, however, the presence of the vertical wall (i.e., no-wall cases in the figures) versus $kD/2$. Here k denotes the wave number and D the diameter of the converter. Also, the Figures 9 and 10 depict the vertical counterpart of the dimensionless exciting forces i.e., $F_z = \frac{\left|F_{D,3}^k\right|}{\frac{\pi \rho g D^2 A}{4}};$ acting on the k

floater, i.e., $k = 1, 3$ compared also with the heave wave loads on the same floater of the array, without the presence of the breakwater (i.e., no-wall cases in the figures), against kD/2.

It can be seen from the Figure 7 that due to the reflected waves from the breakwater the values of the surge exciting forces oscillate around the corresponding values of the same device of the array, without the presence of the wall. Furthermore, it is also evident in the Figure 7 that the horizontal exciting forces are minimizing at kD/2 = 0.80; 1.57; 2.36; ... etc. regardless the shape of the floater or the array configuration. However, this is not the case for the same examined floater and array configuration without the presence of the vertical wall. This phenomenon can be traced back to the interaction effects between the breakwater and the floaters. More specifically, each of these wave numbers (i.e., wave frequencies) correspond to a wavelength equals to a multiple value of the distance between the initial and the image floater, (i.e., $2l_w$).

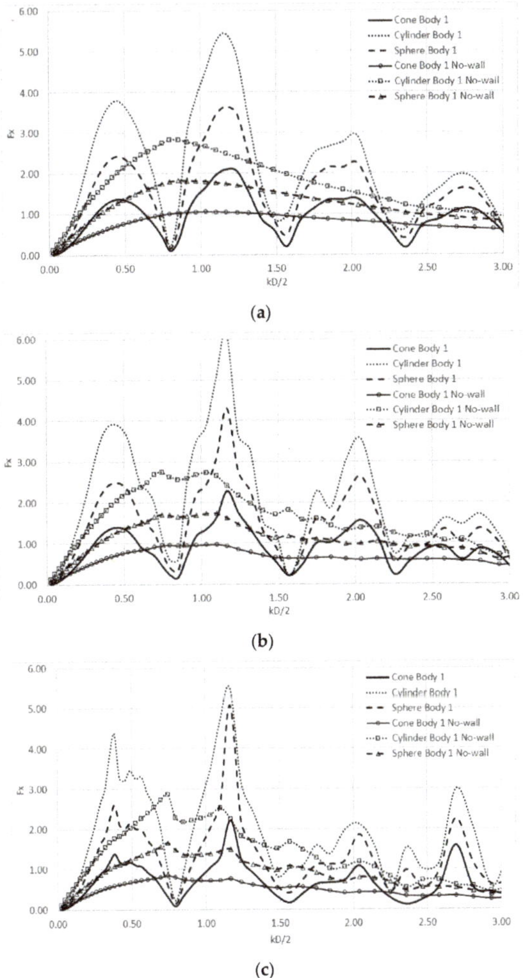

Figure 7. Dimensionless horizontal exciting wave forces acting on the 1st floater of the arrays against kD/2, for the three different examined array configurations. Comparison with the surge exciting wave forces on the same floater of the arrays without the presence of the breakwater: (**a**) parallel arrangement, C_1; (**b**) rectangular arrangement, C_2; (**c**) perpendicular arrangement, C_3.

Moreover, it is also depicted that the horizontal exciting forces on the 1st cylinder of the cylindrical floater array are larger comparing with those on the conical and the semi-spherical floater-array. The reason is the volume of the cylindrical floater which is larger than the volume of the semi-spherical and the conical floater. Finally, it can be seen from the Figure 7b,c, that the scattered waves between the remaining bodies of the perpendicular and rectangular arrangement and the examined floater, create additional peaks (i.e., at kD/2 ≈ 0.6; 1.7; 2.4) on the horizontal forces compared with the corresponding values on the 1st floater of the parallel arrangement. This can be traced back to the position of the floaters of each configuration with respect to the incoming wave train.

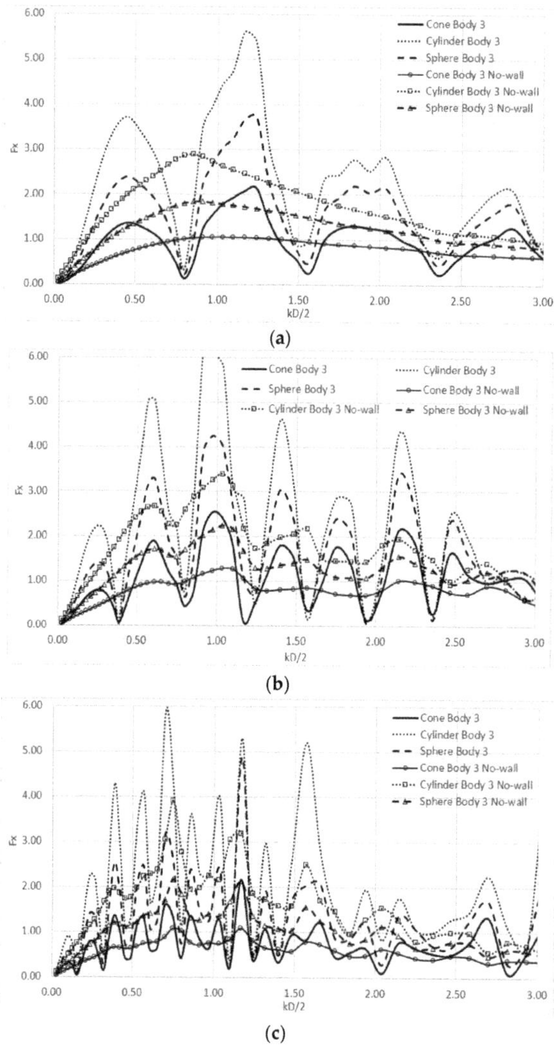

Figure 8. Dimensionless horizontal exciting wave forces acting on the 3rd floater of the arrays plotted against the non-dimensional wave number kD/2, for the three different examined array configurations. Comparison with the surge exciting wave forces on the same floater of the arrays without the presence of the breakwater: (**a**) parallel arrangement, C_1; (**b**) rectangular arrangement, C_2; (**c**) perpendicular arrangement, C_3.

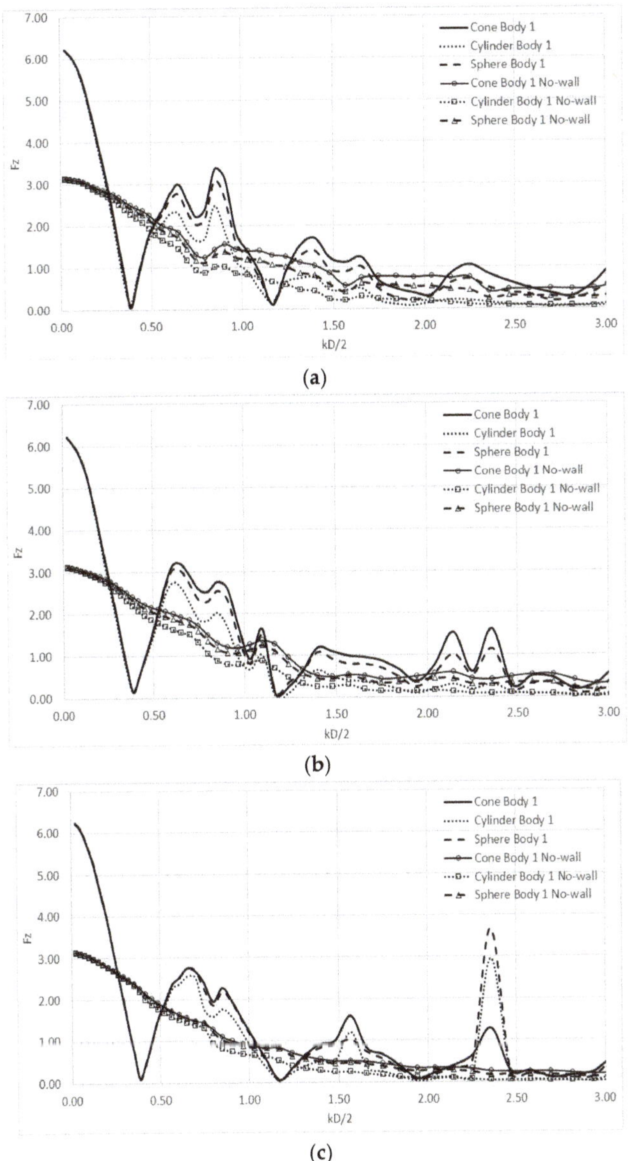

Figure 9. Dimensionless vertical exciting wave forces acting on the 1st floater of the arrays against kD/2, for the three different examined array configurations. Comparison with the heave exciting wave forces on the same floater of the arrays without the presence of the breakwater: (**a**) parallel arrangement, C_1; (**b**) rectangular arrangement, C_2; (**c**) perpendicular arrangement, C_3.

In addition, perpendicular arrangement due to the position of the floaters with respect to the incoming wave. In addition, it can be obtained from Figure 8b,c that the minimization of the horizontal exciting forces does not occur at the same values of kD/2 (i.e., wave frequencies) as in the case of the 1st floater of the corresponding arrays, see Figure 7. It can be derived that at kD/2 = 0.38; 0.8; 1.17; ... etc. for the rectangular array and at kD/2 = 0.15; 0.32; 0.45; ... etc. for the perpendicular array, the wavelength equals to a multiple value of the distance between the initial 3rd floater and its image floater. Also, the sharp peaks observed in the surge exciting forces in Figure 8 can be attributed explicitly to the waves reflected by the wall since they disappear when the wall is removed.

In Figure 9, the heave exciting wave forces on the 1st floater are plotted for every examined array configuration. It is notable that for kD/2 tending to zero, the heave exciting forces on the floater are almost two times larger than the forces on the same floater of the array without the presence of the breakwater. Furthermore, it can be seen that the heave forces minimize at several wave frequencies (i.e., kD/2 = 0.38; 1.17; ... etc.). This behavior does not appear in the cases of the arrays without the existence of the vertical wall. The zeroing of the heave exciting force is due to the interaction phenomena between the floaters and the breakwater and in particular it appears when the distance between the initial 1st floater and its image device equals to a multiple value of the half wave length [47]. The same conclusions can be drawn also from the Figure 10, concerning the double values of the heave exciting forces at kD/2 tending to zero, as well as the minimization of the loads at wave frequencies corresponding to distances, between the initial and image floater, equal to the half of the wave length. Following the conclusions of [48] the duplication of the heave exciting forces when kD/2 tends to zero can be traced back to the fact that a fully wave—reflecting wall of infinite length has been considered here. However, this would not be the case if a finite length breakwater was examined. It can be also seen from the Figures 9 and 10 that, similar to the surge exciting forces, the sharp peaks observed in heave exciting forces can be attributed to the waves reflected by the wall since they disappear when the wall is removed.

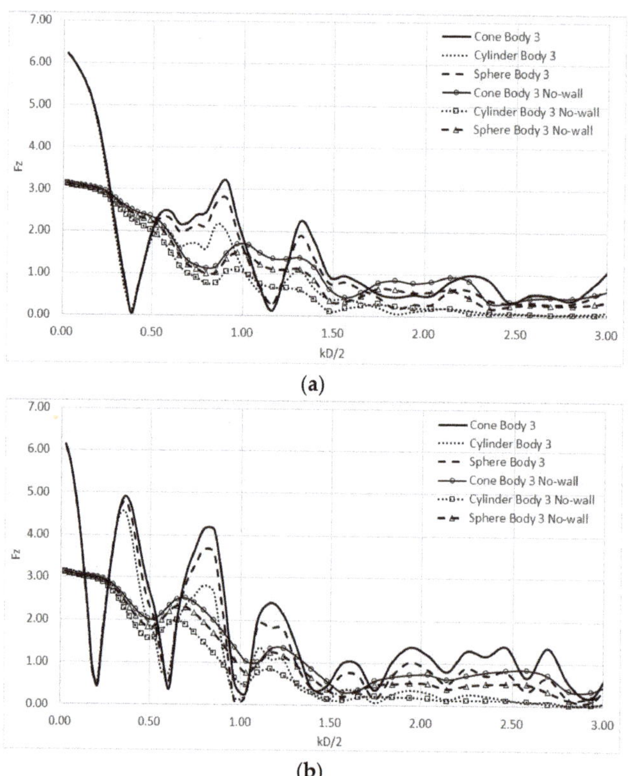

(a)

(b)

Figure 10. *Cont.*

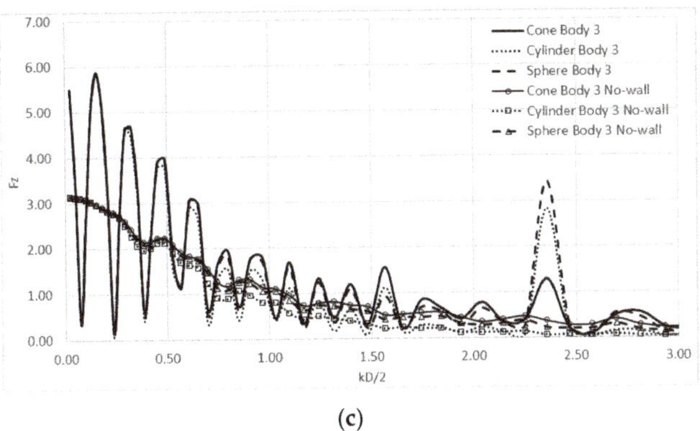

(c)

Figure 10. Dimensionless vertical exciting wave forces acting on the 3rd floater of the arrays against kD/2, for the three different examined array configurations. Comparison with the heave exciting wave forces on the same floater of the arrays without the presence of the breakwater: (**a**) parallel arrangement, C_1; (**b**) rectangular arrangement, C_2; (**c**) perpendicular arrangement, C_3.

Next, the hydrodynamic coefficients of the floaters for the examined array configurations (see Figure 6) are presented against kD/2. More specifically, the hydrodynamic added mass and the damping coefficients as they have been derived by Equation (19) are plotted in a dimensionless form i.e., $A_{ijqp} = \dfrac{a_{i,j}^{q,p}}{\frac{\rho D^3}{8}}$; $B_{ijqp} = \dfrac{\beta_{i,j}^{q,p}}{\frac{\omega \rho D^3}{8}}$; for $i, j \leq 3$; $A_{ijqp} = \dfrac{a_{i,j}^{q,p}}{\frac{\rho D^4}{16}}$; $B_{ijqp} = \dfrac{\beta_{i,j}^{q,p}}{\frac{\omega \rho D^4}{16}}$; for $i \leq 3$ and $j > 3$ or $i > 3$ and $j \leq 3$;

and $A_{ijqp} = \dfrac{a_{i,j}^{q,p}}{\frac{\rho D^5}{32}}$; $B_{ijqp} = \dfrac{\beta_{i,j}^{q,p}}{\frac{\omega \rho D^5}{32}}$; for $i, j > 3$. The results are also compared with the corresponding hydrodynamic characteristics of the same floater of the array, without the presence of the breakwater (i.e., no-wall cases in the presented figures).

In Figure 11, the surge hydrodynamic added mass of the 1st floater due to its forced oscillation in the surge direction is presented for the three array configurations (i.e., parallel; rectangular and perpendicular) and compared with the corresponding values of the added masses of the same floater of the array without the presence of the breakwater. It can be seen that the values of the added mass for the three examined floaters oscillate around the corresponding values referred to the same arrays but without the breakwater. Moreover, it can be observed that for the selected position of the floater in the array (i.e., examined array configuration), a small effect on the added mass coefficient can be reported since the values of the added mass of each floater at each examined array configuration are, in general, quite similar. On the other hand, the type of the floater appears to have a major impact on its added mass. It can be seen that the surge added mass of the cylindrical floater has the larger values compared with the ones of the conical and the semi-spherical floater, due to the cylinder's larger volume. Moreover, it should be also noted that as far as the vertical arrangement is concerned, the interaction phenomena between the cylindrical floaters, oscillating in the surge direction, and the breakwater create a negative added mass at kD/2 = 1.17. However, this is not the case for the conical and the semi-spherical floater.

In Figure 12, the damping coefficient of the 1st floater due to its forced oscillation in the surge direction is presented for the three aforementioned array configurations and floater types (see Figures 1 and 6). The results are also compared with the corresponding damping coefficients referred to the same floater of the arrays, without the presence of the breakwater. It can be seen from the Figure 12, that the breakwater affects also the damping coefficients. The values of the damping coefficients, when the floater is placed in front of the vertical wall, oscillate around those of the same floater without the presence of the wall. Moreover, it can be observed that the damping coefficient maximizes at the same wave frequency (i.e., kD/2 = 1.17) where the surge exciting forces are also maximizing, regardless the examined array configuration. As far as the comparison of the values of the damping coefficient of the examined types of floaters is concerned, it can be seen that due to its larger volume the cylindrical floater is characterized by higher damping coefficients compared with the ones from the conical and the semi-spherical floater.

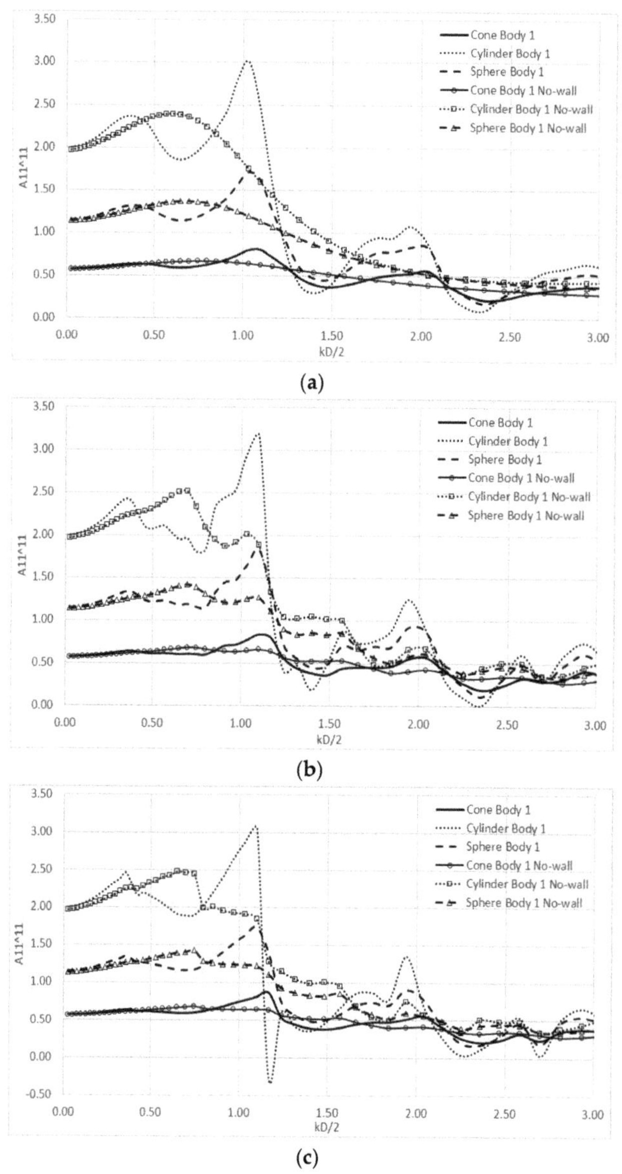

Figure 11. Dimensionless added mass of the 1st floater at the surge direction due to its motion at surge against kD/2, for the three different examined array configurations. Comparison with the corresponding added mass on the same floater of the arrays without the presence of the breakwater: (**a**) parallel arrangement, C_1; (**b**) rectangular arrangement, C_2; (**c**) perpendicular arrangement, C_3.

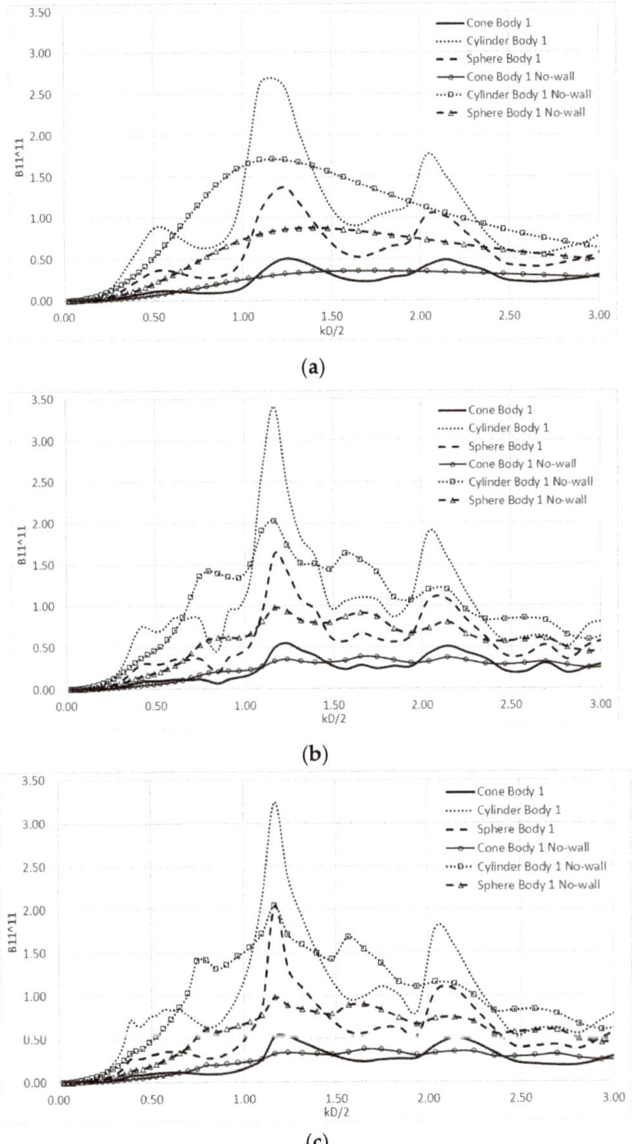

Figure 12. Dimensionless damping coefficient of the 1st floater at the surge direction due to its motion at surge against kD/2, for the three different examined array configurations. Comparison with the corresponding damping coefficient of the same floater of the arrays without the presence of the breakwater: (**a**) parallel arrangement, C_1; (**b**) rectangular arrangement, C_2; (**c**) perpendicular arrangement, C_3.

In Figures 13 and 14, the corresponding surge added mass and damping coefficient of the 3rd floater of the examined array configurations (see Figure 6), due to its surge forced oscillations are presented. Herein, the results are compared also with the corresponding coefficients of the same floater of the array but with absence of the breakwater.

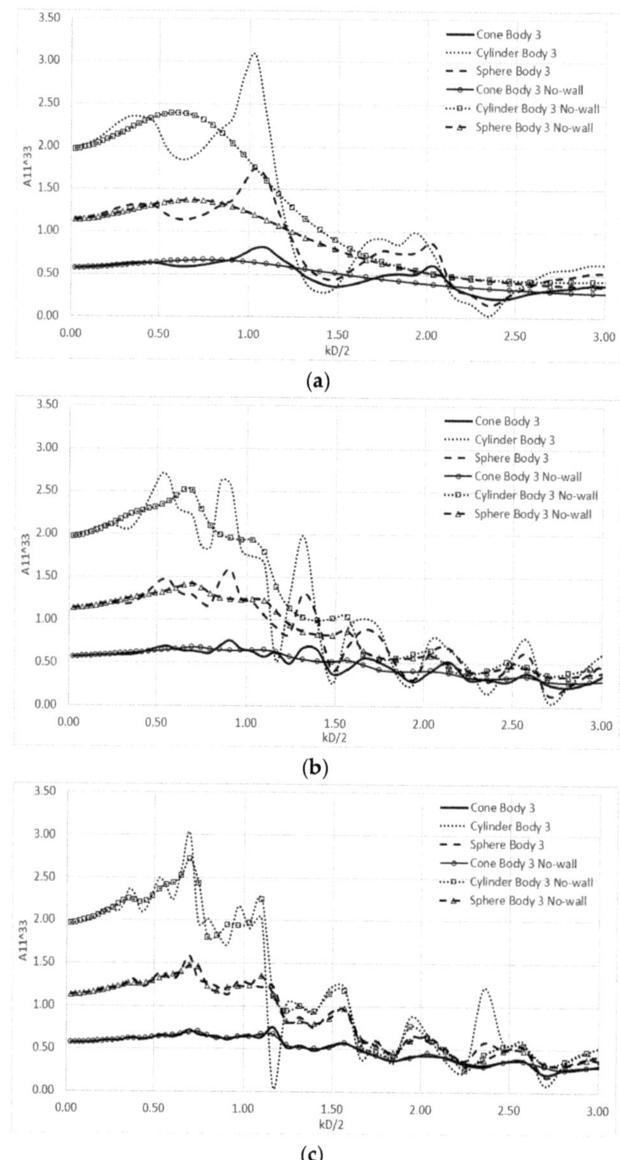

Figure 13. Dimensionless added mass of the 3rd floater at the surge direction due to its motion at surge against kD/2, for the three different examined array configurations. Comparison with the corresponding added mass on the same floater of the arrays without the presence of the breakwater: (**a**) parallel arrangement, C_1; (**b**) rectangular arrangement, C_2; (**c**) perpendicular arrangement, C_3.

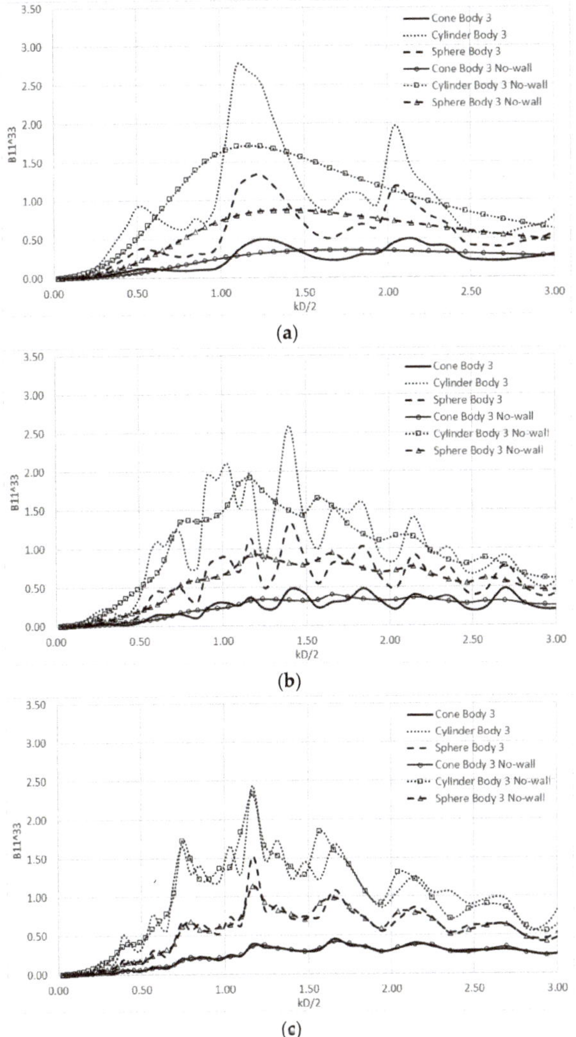

Figure 14. Dimensionless damping coefficient of the 3rd floater at the surge direction due to its motion at surge against kD/2, for the three different examined array configurations. Comparison with the corresponding damping coefficient of the same floater of the arrays without the presence of the breakwater: (**a**) parallel arrangement, C_1; (**b**) rectangular arrangement, C_2; (**c**) perpendicular arrangement, C_3.

Comparing the Figures 11a and 13a it can be seen that the position of each floater in the parallel array configuration does not affect the added mass in surge direction, since both figures (i.e., added mass of the 1st and the 3rd floater) depict very similar results. Furthermore, in these figures the effect of the breakwater is also notable since the values of the added mass of the floater placed in front of a breakwater oscillate around those without the presence of the vertical wall. However, this is not occurring in the cases of the perpendicular and the rectangular arrangements (see Figure 11b,c and Figure 13b,c). The effect of the breakwater on the added mass of the 3rd floater is decreasing for these two arrangements, since the oscillations of the values of the floater's added mass placed in front of the wall, around those of the same floater without the presence of the wall, decrease. This is happening because the 3rd floater is closer to the wall at the parallel array, whereas this distance increases for the rectangular and vertical arrangement. The same conclusion can be drawn for the damping coefficients of the 3rd floater presented in Figure 14. Following the remarks by [49], the arrays of WECs can be divided into broad categories, based on the converters spacing relative to the wavelength. When the ratio of the spacing between the converters of the array and the radius of the WEC has a value larger than the wavelength the interaction phenomena between the converters decrease significantly and each device tends to behave as a single converter. From the image theory presented in this work, the system of the N converters and the breakwater has been simulated as an array of $2N$ converters consisting of the initial and their image virtual devices with respect to the breakwater that are exposed to the action of surface waves (diffraction problem) or forced to move in otherwise calm water (radiation problems) without, however, the presence of the wall. At the examined perpendicular arrangement this ratio is larger than the wavelength for kD/2 > 0.7. However, for lower values of kD/2, where the ratio is tending to the wave length, the interaction phenomena on the initial device due to the motion of its image device are also small. Thus, the breakwater seems to have minor effect on the added mass and on the damping coefficient of the 3rd floater of the array.

Concluding, as far as the surge added mass and damping coefficients are concerned, their values seem to be mainly dictated by the type of the floater (i.e., larger coefficients appear for the cylindrical floater and follow the semi-spherical and the conical floater) and not to a large extend by the floater's array configuration.

In the sequel, the heave hydrodynamic added masses and the heave damping coefficients of the same floaters (i.e., 1st and 3rd, see Figure 6) of the aforementioned array configurations (i.e., parallel, rectangular and vertical arrangement) are presented in Figures 15–18. The Figures 15 and 16 depict the added mass and damping coefficient, respectively, of the 1st floater of the examined array configurations, in heave direction due to its forced oscillation also in heave. Herein, the impact of the floater's type on its hydrodynamic coefficients (i.e., added mass and damping) is notable. The cylindrical floater is characterized by lower values of damping coefficients, compared to the conical and semi-spherical floater, at every examined arrangement. This was not the case for the surge damping coefficient, as being depicted in Figures 12 and 14. Furthermore, it can be observed from the Figures 15 and 16 that the effect of the breakwater is higher for small values of kD/2 (i.e., kD/2 < 1.17). For values of kD/2 larger than 1.17 the values of the added mass and damping coefficients of the floater in front of the breakwater tend to those of the same floater without the presence of the wall. In Figure 16 it can be also seen the doubling of the damping coefficient at values of kD/2 tending to zero. This has been also observed at the heave exciting forces on the floaters in front of the vertical wall.

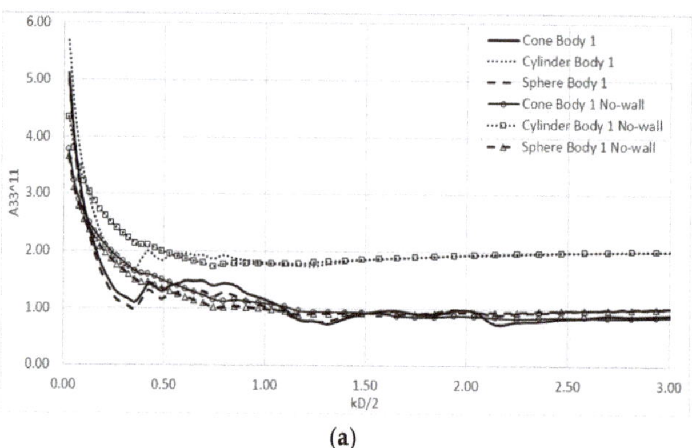

(**a**)

Figure 15. *Cont.*

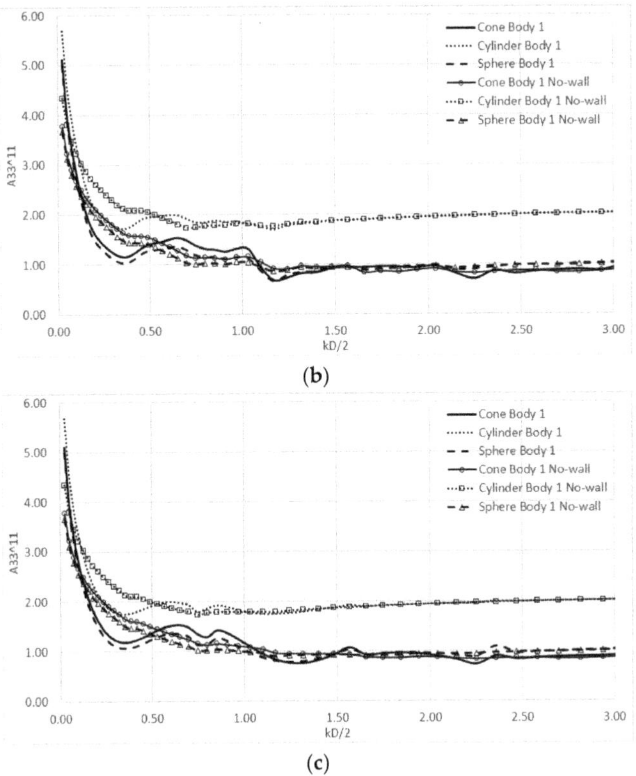

Figure 15. Dimensionless added mass of the 1st floater in the heave direction due to its motion in heave against kD/2, for the three different examined array configurations. Comparison with the corresponding added mass on the same floater of the arrays without the presence of the breakwater: (**a**) parallel arrangement, C_1; (**b**) rectangular arrangement, C_2; (**c**) perpendicular arrangement, C_3.

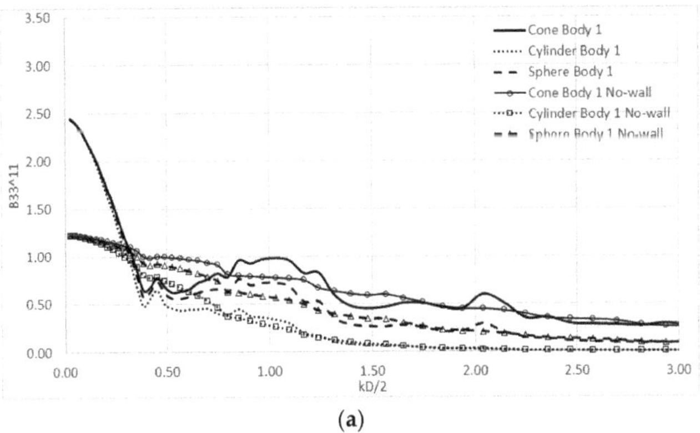

Figure 16. *Cont.*

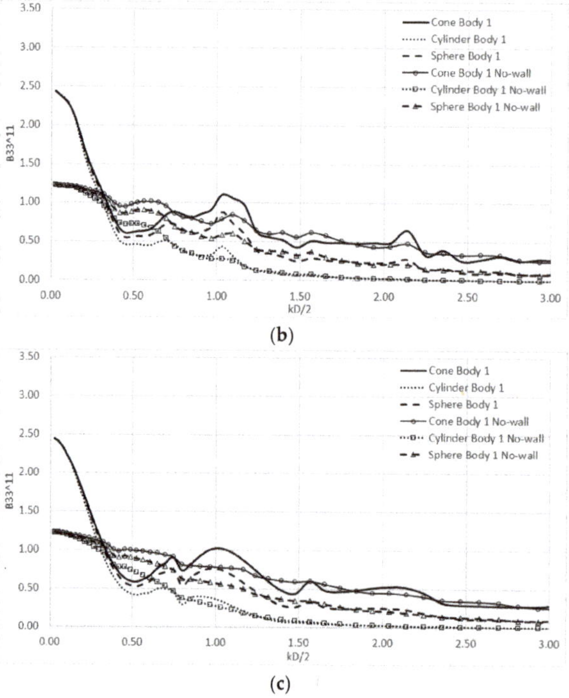

(b)

(c)

Figure 16. Dimensionless damping coefficient of the 1st floater at the heave direction due to its motion at heave against kD/2, for the three different examined array configurations. Comparison with the corresponding damping coefficient of the same floater of the arrays without the presence of the breakwater: (**a**) parallel arrangement, C_1; (**b**) rectangular arrangement, C_2; (**c**) perpendicular arrangement, C_3.

In Figures 17 and 18, the hydrodynamic coefficients of the 3rd floater in heave (i.e., added mass and damping values) are presented for the three examined array configurations in front of the vertical wall and compared with the corresponding values of the same floater of the arrays, without the presence of the vertical wall. Comparing the Figures 15 and 16 with the Figures 17 and 18 it can be seen that the added mass and the damping coefficients of the 3rd floater in the parallel arrangement tend to those of the 1st floater, for the same array configuration.

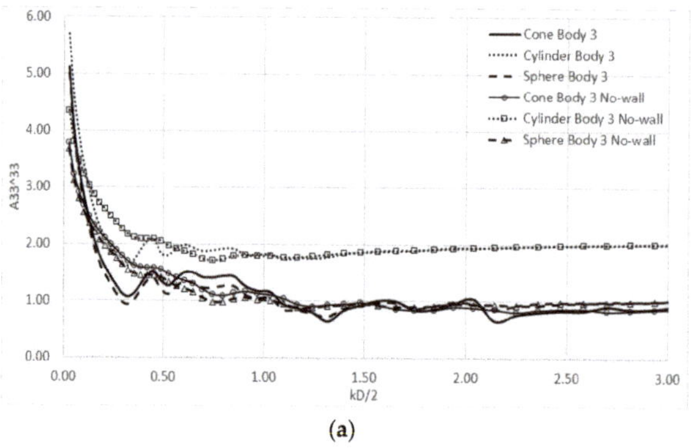

(a)

Figure 17. *Cont.*

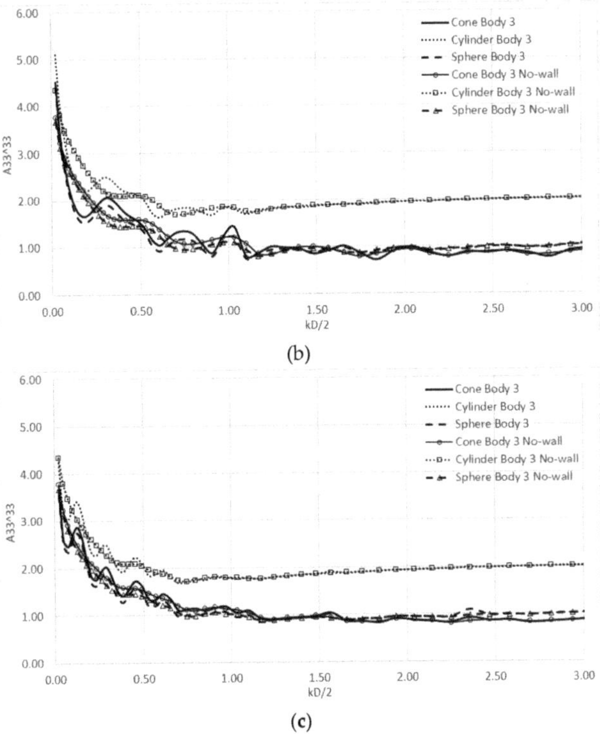

(b)

(c)

Figure 17. Dimensionless added mass of the 3rd floater at the heave direction due to its motion at heave against kD/2, for the three different examined array configurations. Comparison with the corresponding added mass on the same floater of the arrays without the presence of the breakwater: (**a**) parallel arrangement, C_1; (**b**) rectangular arrangement, C_2; (**c**) perpendicular arrangement, C_3.

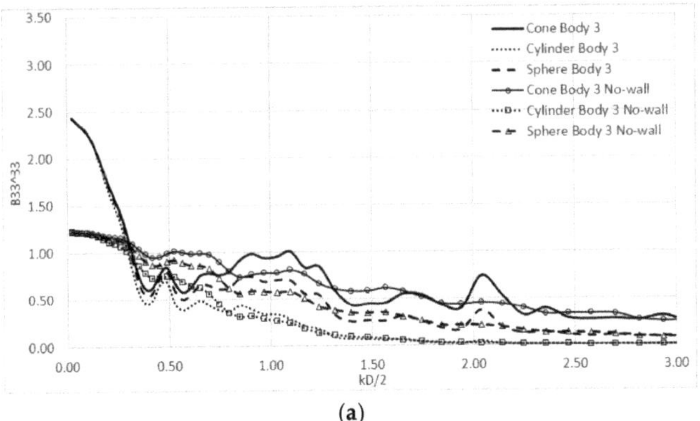

(a)

Figure 18. *Cont.*

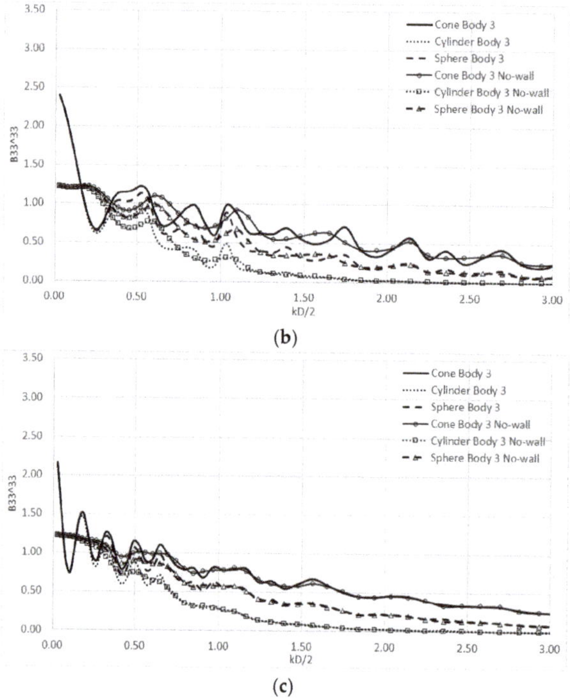

(b)

(c)

Figure 18. Dimensionless damping coefficient of the 3rd floater at the heave direction due to its motion at heave against kD/2, for the three different examined array configurations. Comparison with the corresponding damping coefficient of the same floater of the arrays without the presence of the breakwater: (**a**) parallel arrangement, C_1; (**b**) rectangular arrangement, C_2; (**c**) perpendicular arrangement, C_3.

The same conclusion has been also drawn at the discussion of the Figures 11 and 13. However, this is not the case for the other examined array configurations. Especially, for the perpendicular arrangement, the interaction phenomena between the floaters and the breakwater imply large oscillations of the values of the hydrodynamic coefficients of the 3rd floater, around the ones of the same floater without the presence of the vertical wall. Furthermore, it is notable that the heave hydrodynamic characteristics of the floater is mainly affected by the type of the floater (i.e., cylindrical, conical or semi-spherical) and less by its position in the array.

In the Figures 19 and 20, the hydrodynamic added mass and the damping coefficients in heave are presented for the 3rd floater of the aforementioned types (i.e., cylindrical, conical and semi-spherical) due to the forced heave oscillation of the *j*-th floater ($j = 1, 2, \ldots, 5$), in the perpendicular array case configuration. The latter is selected due to the high interaction phenomena between the members of the array compared with the ones that appear in the parallel and rectangular array. In the Appendix A, these values (i.e., heave hydrodynamic added mass and damping coefficients for the 3rd floater) are also presented in a tabular form for indicative values of wave frequencies to allow more accurate comparisons to be made with other numerical estimates.

Comparing the Figure 19a–c it can be seen that the type of the floater affects mainly the heave added mass due to its own forced heave oscillation. Thus, the higher added mass is applied for the cylindrical floater followed by the added mass from the semi-spherical and conical floater. On the other hand, the added mass on the floater, due to the forced heave motion of the rest bodies of the array, seems not to be affected by the type of the floater, since the values of the hydrodynamic mass of the floaters have similar values. This is also the case for the damping coefficient of the 3rd floater in the heave direction due to the forced oscillation at heave of the *j*-th floater, presented in Figure 20. The contribution of the floater's shape on the damping coefficients of the 3rd floater, due to the forced heave motion of the remaining floaters, seems to be minor. However, contrary to the added mass, the higher values of damping coefficients are presented for the conical and semi-spherical floater, followed by the ones of the cylindrical floater.

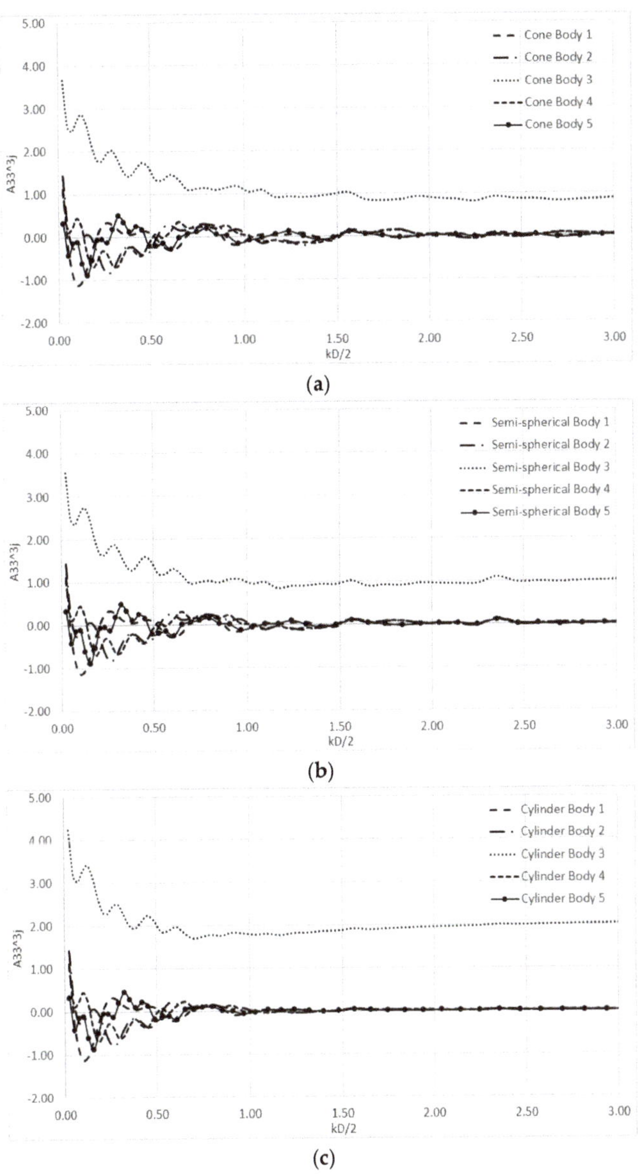

Figure 19. Dimensionless added mass of the 3rd floater in the heave direction due to the forced motion at heave of the *j*-th floater ($j = 1, 2, \ldots, 5$) against kD/2, for the three different examined types of floaters: (**a**) conical floater; (**b**) semi-spherical floater; (**c**) cylindrical floater in the perpendicular arrangement, C_3.

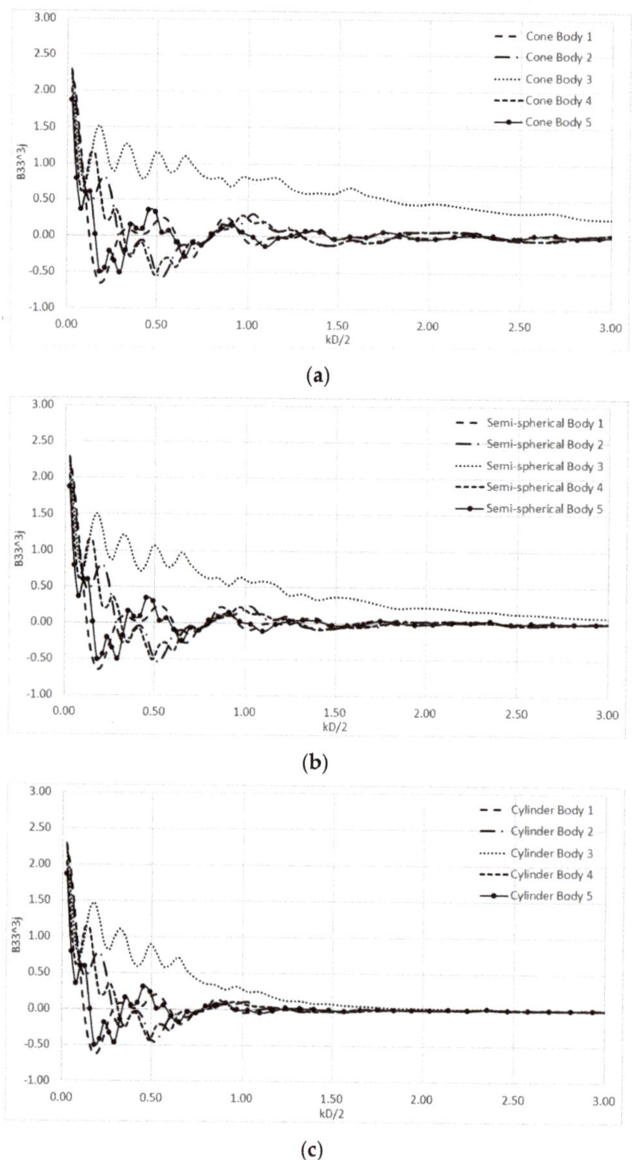

Figure 20. Dimensionless damping coefficient of the 3rd floater at the heave direction due to the forced motion at heave of the j-th floater ($j = 1, 2, \ldots , 5$) against kD/2, for the three different examined types of floaters: (**a**) conical floater; (**b**) semi-spherical floater; (**c**) cylindrical floater in the perpendicular arrangement, C_3.

In the Figures 21 and 22 the hydrodynamic added mass and the damping coefficients in sway (y—direction, see Figure 6) are presented for the 1st floater in the parallel array case configuration, C_1, due to the forced sway oscillation of the j-th floater ($j = 1, 2, \ldots , 5$) in dependence on the floater's geometry, i.e., cylindrical, conical and semi-spherical. It can be seen from the Figures 21 and 22 that the type of the floater affects its hydrodynamic characteristics not only due to its own forced sway oscillation but also to the forced oscillation in the sway direction of the rest of the floaters. Thus, due to its larger volume the cylindrical floater is characterized by higher hydrodynamic characteristics compared with the ones from the conical and the semi-spherical floater. Indicative theoretical results for the cylindrical floater case, i.e., Figures 21c and 22c, are given in the Appendix B, (i.e., sway hydrodynamic added mass and damping coefficients for the 1st floater) in tabular form to allow more accurate comparisons to be made with other numerical estimates.

In the Figure 23 the hydrodynamic added mass and the damping coefficient in sway (y—direction, Figure 6) are presented indicatively for the 1st conical floater due to the forced surge oscillation (x—direction) of the j-th conical floater ($j = 1, 2, \ldots , 5$), in the parallel array case configuration, C_1. The results are compared with the corresponding values of the 1st conical floater of the array without the presence of the breakwater. Due to the examined arrangement (i.e., parallel) these values (i.e., hydrodynamic added mass and damping coefficient in sway for the 1st conical floater due to the forced surge oscillation of the j-th conical floater) tend to zero at every wave frequency, when the vertical wall is absent. Nevertheless, this is not the case for the array placed in front of the breakwater, since it is evident from the figures that the examined hydrodynamic characteristics of the floater appear no zero values. Indicative theoretical results for the hydrodynamic interaction coefficients concerning the 1st conical floater, see Figure 23a,b, are given in the Appendix C, in tabular form to allow more accurate comparisons to be made with other numerical estimates.

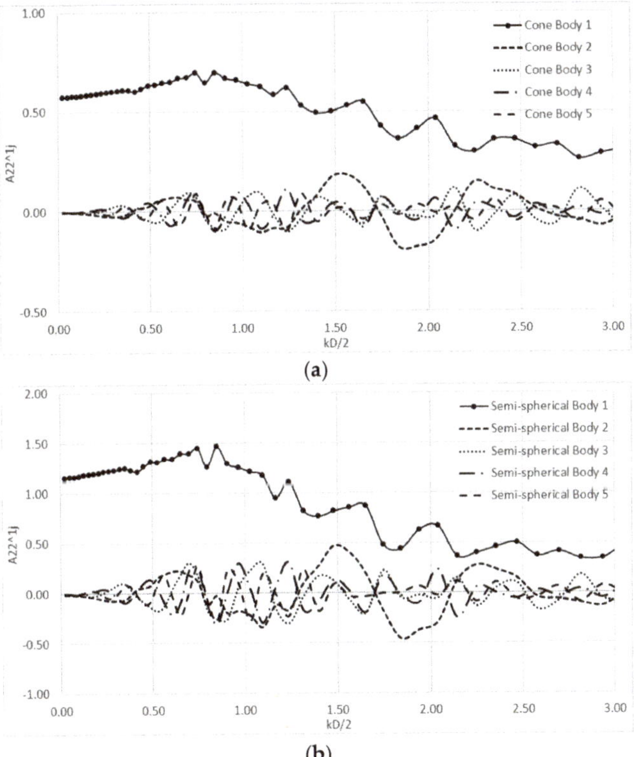

(a)

(b)

Figure 21. *Cont.*

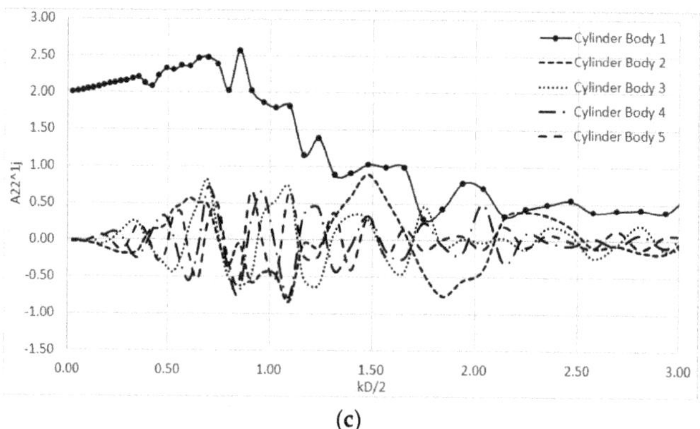

(**c**)

Figure 21. Dimensionless added mass of the 1st floater in the sway direction due to the forced motion in sway of the j-th floater ($j = 1, 2, \dots , 5$) against kD/2, for the three different examined types of floaters: (**a**) conical floater; (**b**) semi-spherical floater; (**c**) cylindrical floater in the parallel array case configuration, C_1.

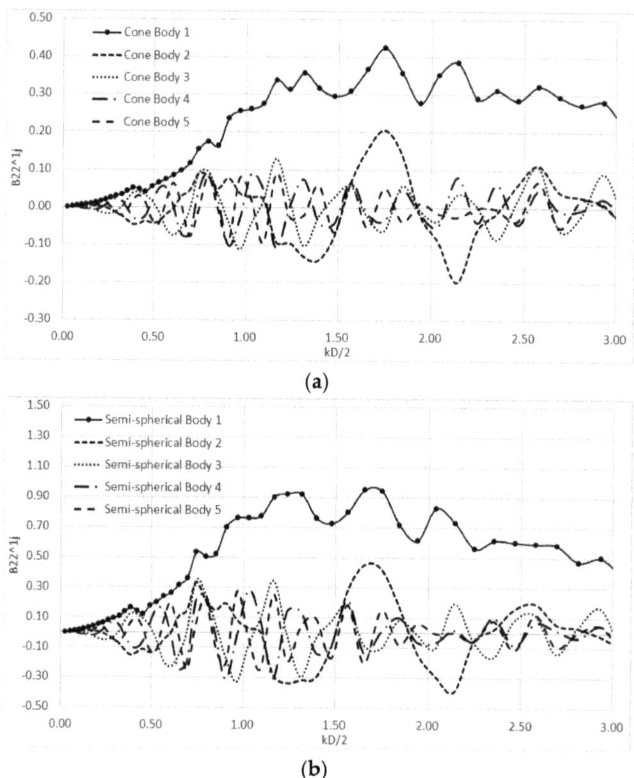

(**a**)

(**b**)

Figure 22. *Cont.*

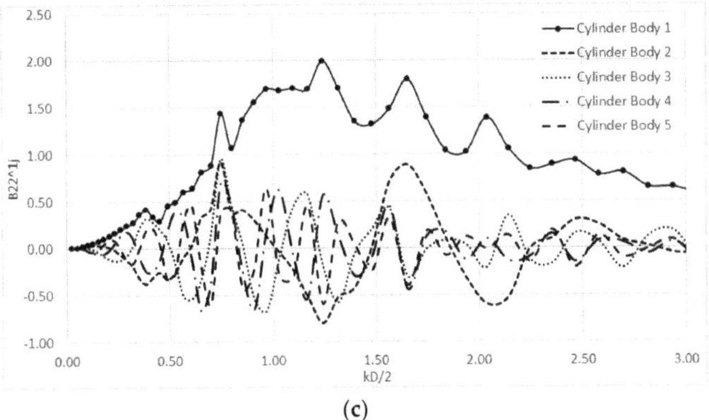

(c)

Figure 22. Dimensionless damping coefficient of the 1st floater at the sway direction due to the forced motion in sway of the *j*-th floater ($j = 1, 2, \ldots , 5$) against kD/2, for the three different examined types of floaters: (**a**) conical floater; (**b**) semi-spherical floater; (**c**) cylindrical floater in the parallel array case configuration, C_1.

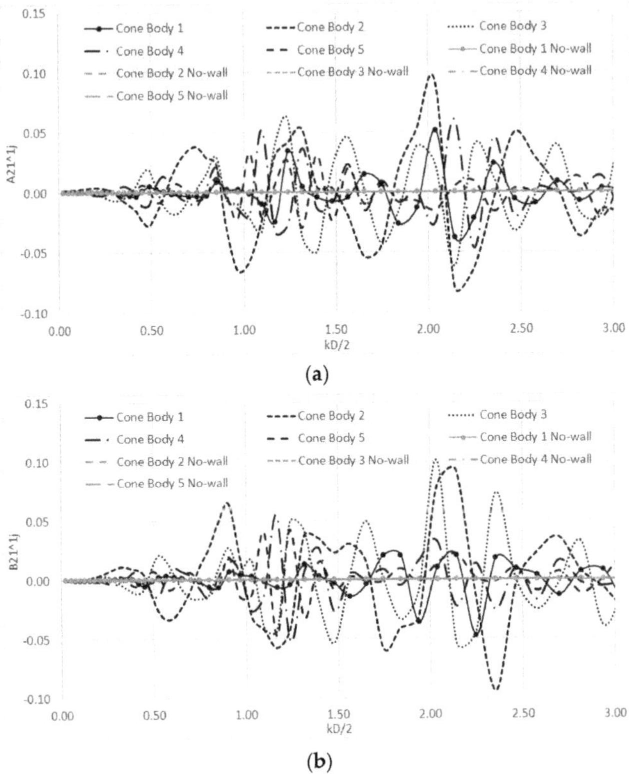

Figure 23. Dimensionless added mass (**a**) and damping coefficient (**b**) of the 1st conical floater in the sway direction due to the forced motion in surge of the *j*-th conical floater ($j = 1, 2, \ldots , 5$) against kD/2, for the parallel arrangement, C_1.

Next, indicative results concerning the absorbed wave power by the three examined arrangements (i.e., parallel, rectangular and perpendicular to the wall) for the case of the cylindrical floater-array are given. The cylindrical floaters of the array are assumed to oscillate at the heave direction, as heaving WECs. The power absorbed by each floater equals to [50]:

$$p_q = 0.5 b_{PTO} \omega^2 \left| x_{30}^q \right|^2 \tag{20}$$

Here, x_{30}^q is the complex amplitude of q floater's motion in the vertical direction obtained from the solution of the linear system of motion equations (see Appendix D); ω is the wave frequency and b_{PTO} represents the damping coefficient that originates from the Power Take Off (PTO) mechanism, modeled as a linear damping system actuated from the heave motion of the floater. In the present analysis all the examined cylindrical floaters are considered to have the same PTO characteristics, which for the sake of presenting some initial numerical results are assumed, according to [50], equal to the heave radiation damping of the isolated cylindrical floater at its heave natural frequency. Considering the draught and the radius of the floaters equal to 5 m and the mass of each floater equals to 402.5 t, the heave natural frequency of the floater equals to 1.088 rad/s, leading to $b_{PTO} = 73.88$ kNm/s.

In the Figure 24 the absorbed power by the first cylindrical floater (see Figure 6) of each examined arrangement is presented against the corresponding results of the same array arrangements without the presence of the breakwater. It can be observed that the wall influences the absorbed wave power by the floater. More specifically, near the natural frequency of the floater (i.e., kD/2 = 0.68) the presence of the wall causes a significant increase of the floater's absorbed power value compared to its no wall counterpart. This is in line with the expected influence of the vertical wall, which actually offers to the WEC the reflected wave for harvesting. This influence is evident in the low frequency regime of e.g., Figure 9 (wave frequencies up to approximately 0.25 rad/s), where the vertical exciting wave forces on the first floater of the examined array configuration are depicted. However, for higher than 0.25 rad/s wave frequencies the wall's influence is affected by the hydrodynamic interactions between the floaters and the wall, depending on the specific geometrical arrangement and the distances l_w. These interactions can minimize the excitation forces and reduce the absorbed power. Moreover, it has to be stressed that the wave power absorption by a floater in front of a wall and its comparison with the no wall case is affected by the PTO damping coefficient which has to be properly selected to accommodate the importance of the hydrodynamic interference effects. It is therefore evident that the effect of the vertical wall on the WECs efficiency should be further examined for different PTO damping coefficients, distances between the wall and the floaters and various angles of wave propagation and in the entire range of the wave frequencies of interest, in order to depict accurately the influence of the wall to the WECs efficiency.

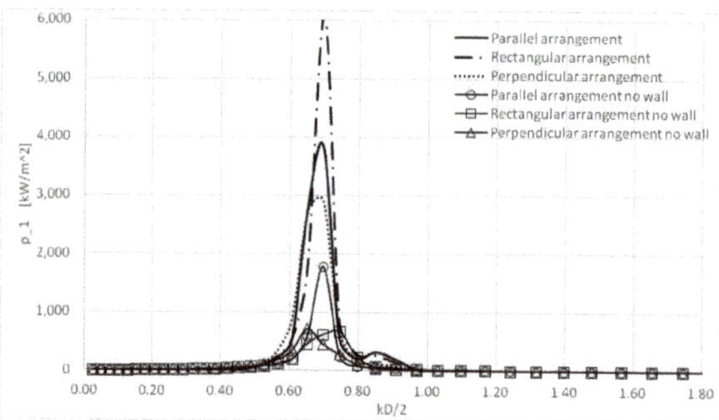

Figure 24. Absorbed wave power by the 1st cylindrical floater against kD/2, for the parallel, C_1, rectangular, C_2; and perpendicular, C_3 arrangement. The results are also compared with the absorbed power of the same floater of the arrangements without the presence of the wall.

5. Conclusions

This study dealt with the determination of the hydrodynamic loads on an array of floaters in front of a vertical breakwater of infinite length. The image method has been applied to simulate the effect of the breakwater on the array and the multiple scattering approach has been used to evaluate the interaction phenomena between the WECs.

Three different types of floaters have been studied, i.e., a cylindrical; a conical and a semi-spherical floater, as well as three different types of array configurations in front of the vertical wall have been investigated, i.e., a parallel, a perpendicular and a rectangular array. Based on the theoretical computations shown and discussed in the dedicated sections, the main findings of the present research contribution concern the effect of the breakwater on the exciting wave forces and the hydrodynamic coefficients of the floaters, at every examined configuration, which should not be neglected when designing a WEC array in front of a vertical wall. This effect, the significance of which is depending on the distances between the floaters and the wall, causes an increase or decrease of the values of the exciting loads and the hydrodynamic coefficients of the floaters, at specific wave frequencies. On top of that, it is shown that the type of the floater dictates by a large expense the exciting wave loads and its hydrodynamic characteristics compared to the effect that the different examined array configuration may have on these hydrodynamic parameters. The presented figures showed that the cylindrical floater is characterized by higher values of surge exciting forces, as well as, surge hydrodynamic added mass and damping coefficients, at every array configuration, compared to those of the conical and semi-spherical floaters. However, this is not the case for the heave damping coefficients of the cylindrical floater, which attain lower values compared to the corresponding ones of the other examined floaters.

Concluding, the effect of the breakwater on the exciting wave loads and hydrodynamic coefficients of a floater, when the latter is a part of an array placed in front of the wall, increases or decreases in dependence to the ratio of the distance between the wall and the breakwater and the radius of the examined floater. However, neither the existence of the breakwater nor the arrangement of the array with respect to the incoming wave seem to have a greater influence on the values of the hydrodynamic forces and coefficients of a floater than its geometrical characteristics. Nevertheless, the present research will be continued further by determining in detail the power efficiency of the array and the q-factor [39] of the system, as well as their interaction with the presence of the breakwater, the shape of the floaters and their position in the array towards the incoming wave.

Author Contributions: Conceptualization, S.A.M. and G.M.K.; methodology, S.A.M., D.N.K., G.M.K.; software, S.A.M., D.N.K.; validation, D.N.K., G.M.K.; formal analysis, G.M.K.; investigation, D.N.K., G.M.K. and S.A.M.; data curation, D.N.K., G.M.K.; writing—original draft preparation, D.N.K.; writing—review and editing, D.N.K., S.A.M., G.M.K.; supervision, S.A.M., G.M.K.; project administration, S.A.M. All authors have read and agreed to the published version of the manuscript.

Funding: This research received no external funding.

Conflicts of Interest: The authors declare no conflict of interest.

Nomenclature

N	Number of vertical axisymmetric floaters
d	Water depth
ω	Wave frequency
A	Wave amplitude
θ	Angle of wave propagation with respect to the positive x-axis
D	Outer diameter of the examined floater
h	Distance of the floater's bottom from the sea bed
l_W	Distance between the center of the closest to the wall floater and the breakwater
l_b	Distance between adjacent floaters

φ^q	Potential function describing the fluid flow around the q floater, i.e., $\Phi\left(r_q, \theta_q, z_q; t\right) = Re\left[\varphi^q\left(r_q, \theta_q, z_q\right)\right]$
φ_0	Velocity potential of the undisturbed incident wave
φ_s^q	Velocity potential of the scattered wave field around the q floater
φ_j^q	Velocity potential of the radiated wave field around the q floater
φ_D^q	Diffraction velocity potential around the q floater
J_m	The m-th order Bessel function of the first kind
k	Wave number
$\left(l_{0q}, \theta_{0q}\right)$	The polar coordinates of the q floater center relative to the origin of the global co-ordinate system
Z_0, Z_n	Orthonormal functions
$\dot{x}_{j0}^q$	The complex velocity amplitude of q floater's motion in the j-th direction
I	The infinite ring element around the q floater
III_p	The p-th ring element, $p = 1, \ldots, L$, below the q floater
α_p	The radius of the p-th ring element of type III
h_p	Distance of the upper surface of the p-th ring element of type III from the sea bottom
H_m	The m-th order Hankel function of first kind
K_m	The m-th order modified Bessel function of second kind
$F_{k,mn}^q$ F_{k,mn_p}^q F_{k,mn_p}^{*q}	The unknown Fourier coefficients
ϵ_{n_p}	The Neumann's symbol, $p = 1, \ldots, L$
a_n	The roots of the transcendental equation: $\omega^2 + g a_n \tan(a_n d) = 0$
ρ	Sea water density
$F_{D,i}^q$	The exciting wave forces on the q floater in the i-th direction
F_{ij}^{qp}	The hydrodynamic reaction forces acting on the q floater in the i-th direction, due to the forced oscillation of the p floater in the j-th direction
S_q	The mean wetted surface of the q floater
n_i	The generalized normal components
r	The position vector of a point on the wetted surface S_q with respect to the reference co-ordinate system of q floater
$a_{i,j}^{q,p}$	The added mass of the q floater in i-th direction due to the forced oscillation of the p floater in the j-th direction
$\beta_{i,j}^{q,p}$	The damping coefficient of the q floater in i-th direction due to the forced oscillation of the p floater in the j-th direction
b_{PTO}	The PTO damping coefficient
p_q	Absorbed wave power by the q floater

Appendix A

In the Tables A1–A3 the hydrodynamic characteristics in heave of the 3rd floater of an array of five floaters in front of a vertical breakwater (arrangement C_3, Figure 6) due to the heave motion of the p-th body in the array $(p = 1, \ldots, 5)$, i.e., $A_{333p} = \dfrac{a_{3,3}^{3,p}}{\frac{\rho D^3}{8}}$; $B_{333p} = \dfrac{\beta_{3,3}^{3,p}}{\frac{u \rho D^3}{8}}$, (see Equation (19)), are presented against kD/2, for the examined arrays of conical, semi-spherical and cylindrical floaters, respectively.

Table A1. Hydrodynamic characteristics of the 3rd floater of the array of five conical floaters.

kD/2	A33^31	B33^31	A33^32	B33^32	A33^33	B33^33	A33^34	B33^34	A33^35	B33^35
0.0760	−0.2360	1.1028	−0.8603	1.2179	2.4640	0.8949	−0.1510	0.3629	0.2158	0.6327
0.1805	−0.0252	0.7306	−0.3496	−0.6498	2.1878	1.5191	−0.5434	−0.5031	−0.7096	0.6787
0.2928	−0.7786	−0.0826	0.2841	0.0017	2.0292	1.0917	0.1770	−0.5122	−0.6660	0.2569
0.3536	−0.2798	−0.2549	0.0789	0.0752	1.5308	1.2013	0.3457	0.1524	−0.4136	−0.2889
0.4894	−0.3789	−0.4587	0.1129	0.1740	1.6349	1.1434	−0.2070	0.3310	−0.1616	−0.5393
0.5665	0.2184	−0.4510	−0.2133	0.1988	1.3301	0.8888	−0.1542	0.0658	−0.0215	−0.3010
0.6515	0.1568	−0.1270	−0.2278	−0.2506	1.3399	1.1049	−0.0274	−0.2827	0.3463	−0.2111
0.7459	0.2740	−0.1072	0.2078	−0.1297	1.1247	0.8422	0.1125	−0.1327	0.2257	−0.1059
0.8511	0.2465	0.1522	0.1636	0.2200	1.0931	0.8060	0.0502	0.1023	0.1927	0.0828
0.9683	0.1613	0.1726	−0.1883	−0.0104	1.1807	0.8200	−0.1564	0.0511	0.0755	0.2864
1.0984	−0.1636	0.1469	0.0456	−0.0476	1.0931	0.7917	−0.0254	−0.1468	−0.1188	0.1967
1.3980	−0.0915	−0.1084	−0.0460	0.1049	0.9388	0.6009	−0.0733	0.0654	−0.1622	−0.1120
1.5668	0.1309	−0.0473	0.1312	0.0015	1.0226	0.6763	0.0959	0.0004	0.1154	−0.0474
1.7476	0.0903	−0.0753	−0.0075	0.0715	0.8328	0.5286	0.0126	0.0632	0.1197	−0.0466
1.9397	0.0314	0.0606	−0.0039	−0.0147	0.9211	0.4426	−0.0080	−0.0302	0.0322	0.0677
2.1428	−0.0420	0.0704	0.0427	−0.0067	0.8566	0.4343	0.0330	0.0079	−0.0113	0.0764
2.3564	0.0234	−0.0184	0.0627	−0.0189	0.9101	0.3617	0.0254	0.0180	0.0049	−0.0015
2.8148	0.0482	−0.0051	−0.0109	−0.0057	0.8368	0.2777	−0.0160	−0.0170	0.0489	0.0003

Table A2. Hydrodynamic characteristics of the 3rd floater of the array of five semi-spherical floaters.

kD/2.	A33^31	B33^31	A33^32	B33^32	A33^33	B33^33	A33^34	B33^34	A33^35	B33^35
0.0760	−0.8635	1.2148	−0.2368	1.1000	2.3526	0.8922	0.2178	0.6306	−0.1497	0.3611
0.1805	−0.3377	−0.6459	−0.0255	0.7255	2.0528	1.5033	−0.7073	0.6624	−0.5302	−0.5045
0.2928	0.2713	0.0112	−0.7435	−0.0973	1.8795	1.0634	−0.6481	0.2397	0.1851	−0.4980
0.3536	0.0697	0.0699	−0.2516	−0.2337	1.3792	1.1262	−0.3762	−0.2743	0.3186	0.1590
0.4894	0.1053	0.1708	−0.3529	−0.4431	1.4818	1.0611	−0.1367	−0.5065	−0.2121	0.2974
0.5665	−0.2086	0.1730	0.2169	−0.3894	1.1993	0.7768	−0.0269	−0.2731	−0.1417	0.0625
0.6515	−0.1809	−0.2476	0.1228	−0.1022	1.1718	0.9735	0.3082	−0.1472	0.0078	−0.2533
0.7459	0.1729	−0.0814	0.2182	−0.0751	0.9993	0.6822	0.1856	−0.0862	0.1099	−0.1063
0.8511	0.1182	0.2058	0.1870	0.1343	0.9866	0.6190	0.1473	0.0605	0.0313	0.0802
0.9683	−0.1454	−0.0338	0.1254	0.1400	1.0568	0.6282	0.0262	0.2212	−0.1289	0.0170
1.0984	0.0373	−0.0214	−0.1192	0.0709	0.9910	0.5761	−0.0992	0.1234	0.0071	−0.1188
1.3980	−0.0485	0.0664	−0.0432	−0.0657	0.9364	0.3272	−0.0918	−0.0885	−0.0619	0.0327
1.5668	0.1177	−0.0224	0.1128	−0.0457	1.0162	0.3594	0.1023	−0.0503	0.0942	−0.0324
1.7476	−0.0175	0.0401	0.0557	−0.0414	0.9123	0.2609	0.0660	−0.0130	−0.0058	0.0377
1.9397	−0.0006	−0.0068	0.0105	0.0251	0.9558	0.2206	0.0105	0.0300	−0.0018	−0.0174
2.1428	0.0254	0.0013	−0.0227	0.0273	0.9354	0.2011	−0.0070	0.0384	0.0186	0.0096
2.3564	0.1268	0.0071	0.1106	0.0097	1.0873	0.1657	0.0914	0.0234	0.0862	0.0364
2.8148	−0.0019	−0.0036	0.0110	0.0014	0.9984	0.0946	0.0089	−0.0001	−0.0010	−0.0021

Table A3. Hydrodynamic characteristics of the 3rd floater of the array of five cylindrical floaters.

kD/2	A33^31	B33^31	A33^32	B33^32	A33^33	B33^33	A33^34	B33^34	A33^35	B33^35
0.0760	−0.8664	1.2090	−0.2374	1.0945	3.0262	0.8876	0.2201	0.6269	−0.1477	0.3583
0.1805	−0.3195	−0.6352	−0.0257	0.7127	2.6931	1.4694	−0.6980	0.6354	−0.5085	−0.5018
0.2928	0.2484	0.0224	−0.6812	−0.1127	2.4978	1.0035	−0.6103	0.2122	0.1914	−0.4677
0.3536	0.0568	0.0609	−0.2102	−0.2000	2.0120	0.9968	−0.3184	−0.2468	0.2732	0.1608
0.4894	0.0873	0.1554	−0.2938	−0.3956	2.1036	0.8971	−0.0964	−0.4327	−0.2007	0.2357
0.5665	−0.1818	0.1275	0.1911	−0.2858	1.8794	0.5890	−0.0282	−0.2209	−0.1173	0.0530
0.6515	−0.1067	−0.2072	0.0754	−0.0699	1.8190	0.7170	0.2225	−0.0680	0.0404	−0.1837
0.7459	0.1062	−0.0356	0.1313	−0.0411	1.7336	0.4294	0.1208	−0.0532	0.0868	−0.0606
0.8511	0.0511	0.1419	0.0973	0.0879	1.7539	0.3439	0.0834	0.0306	0.0146	0.0443
0.9683	−0.0712	−0.0342	0.0641	0.0800	1.8073	0.3221	−0.0093	0.1074	−0.0671	−0.0080
1.0984	0.0145	−0.0034	−0.0436	0.0155	1.7947	0.2482	−0.0459	0.0415	0.0196	−0.0510
1.3980	−0.0206	0.0164	−0.0085	−0.0166	1.8519	0.0885	−0.0188	−0.0268	−0.0214	0.0041
1.5668	0.0423	−0.0165	0.0399	−0.0211	1.9072	0.0634	0.0316	−0.0260	0.0327	−0.0210
1.7476	−0.0050	0.0059	0.0115	−0.0063	1.9025	0.0426	0.0096	0.0011	−0.0027	0.0057
1.9397	0.0000	−0.0007	0.0013	0.0024	1.9347	0.0278	0.0065	0.0063	0.0003	−0.0019
2.1428	0.0027	0.0009	−0.0018	0.0020	1.9542	0.0177	−0.0010	0.0035	0.0018	0.0017
2.3564	0.0176	0.0108	0.0168	0.0110	1.9931	0.0212	0.0126	0.0118	0.0125	0.0120
2.8148	0.0001	−0.0002	0.0007	0.0001	2.0082	0.0032	0.0010	−0.0003	0.0002	0.0003

Appendix B

In the Table A4 the hydrodynamic characteristics in sway of the 1st cylindrical floater of an array of five cylindrical floaters in front of a vertical breakwater (arrangement C_1, Figure 6) due to the sway motion of the *p*-th body in the array ($p = 1, \dots, 5$), i.e., $A_{22\Gamma1p} = \dfrac{a_{2,2}^{1,p}}{\frac{\rho D^3}{8}}$; $B_{22\Gamma1p} = \dfrac{\beta_{2,2}^{1,p}}{\frac{\omega \rho D^3}{8}}$, (see Equation (19)), are presented against kD/2.

Table A4. Hydrodynamic characteristics of the 1st floater of the array of five cylindrical floaters.

kD/2	A22^11	B22^11	A22^12	B22^12	A22^13	B22^13	A22^14	B22^14	A22^15	B22^15
0.0760	2.0240	0.0183	−0.0333	0.0156	−0.0229	0.0089	−0.0208	0.0000	−0.0165	−0.0084
0.1805	2.0946	0.0988	−0.1033	0.0271	−0.0532	−0.0743	0.0491	−0.0597	0.0608	0.0351
0.2928	2.1574	0.2369	−0.1880	−0.1245	0.1852	−0.0818	−0.0532	0.1774	−0.0791	−0.1307
0.3536	2.2015	0.3541	−0.1477	−0.3186	0.2236	0.2107	−0.2359	−0.0907	0.2053	−0.0131
0.4894	2.3223	0.4508	0.2054	−0.3321	−0.3508	0.0840	0.3204	0.1705	−0.1637	−0.3285
0.5665	2.3637	0.6022	0.4733	−0.0971	−0.2355	−0.5107	−0.3272	0.3323	0.3707	0.2725
0.6515	2.4707	0.8085	0.5054	0.3262	0.5083	−0.3324	−0.1108	−0.6708	−0.5270	−0.2038
0.7459	2.3836	1.4354	0.0344	0.9222	0.1344	0.9514	0.3748	0.6947	0.4932	0.3823
0.8511	2.5715	1.3667	−0.0572	0.4063	−0.6186	0.0419	−0.7277	−0.2513	−0.5758	−0.4337
0.9683	1.8682	1.6947	−0.4459	0.1238	0.3490	−0.6673	0.6573	0.1634	−0.0601	0.6370
1.0984	1.8146	1.6999	−0.8365	−0.2386	0.7164	0.4411	−0.7716	−0.0823	0.6550	−0.3169
1.3980	0.9078	1.3463	0.5921	−0.4201	0.3349	−0.0537	−0.0863	−0.2994	−0.4034	0.0289
1.5668	0.9926	1.4841	0.5586	0.7083	−0.2482	0.4221	−0.1401	0.4517	−0.2159	0.3162
1.7476	0.2887	1.3876	−0.4381	0.5763	0.4545	−0.0133	0.2373	0.1227	−0.1345	0.1795
1.9397	0.7768	1.0233	−0.5525	−0.3183	0.0166	0.0321	−0.1709	−0.1282	0.0572	0.1157
2.1428	0.3299	1.0565	0.2992	−0.5280	0.0286	0.3425	−0.3123	−0.1373	0.1857	0.1297
2.3564	0.4857	0.8862	0.3346	0.1012	0.1713	−0.1660	−0.0121	0.1464	0.0771	0.1877
2.8148	0.4188	0.6489	−0.1544	0.0188	0.2062	0.1173	0.0848	−0.0616	−0.1132	−0.0147

Appendix C

In the Table A5 the hydrodynamic characteristics in sway of the 1st conical floater of an array of five conical floaters in front of a vertical breakwater (arrangement C_1, Figure 6) due to the surge motion of the *p*-th body in the array ($p = 1, \dots, 5$), i.e., $A_{2\Gamma1p} = \dfrac{a_{2,1}^{1,p}}{\frac{\rho D^3}{8}}$; $B_{2\Gamma1p} = \dfrac{\beta_{2,1}^{1,p}}{\frac{\omega \rho D^3}{8}}$, (see Equation (19)), are presented against kD/2.

Table A5. Hydrodynamic characteristics of the 1st floater of the array of five conical floaters.

kD/2	A21^11	B21^11	A21^12	B21^12	A21^13	B21^13	A21^14	B21^14	A21^15	B21^15
0.0760	0.0000	0.0000	0.0025	0.0001	0.0010	0.0002	0.0005	0.0002	0.0002	0.0002
0.1805	0.0002	0.0000	0.0044	0.0024	0.0006	0.0027	−0.0014	0.0010	−0.0010	−0.0006
0.2928	0.0000	0.0006	0.0008	0.0099	−0.0062	−0.0002	0.0014	−0.0034	0.0018	0.0024
0.3536	−0.0007	0.0011	−0.0069	0.0110	−0.0043	−0.0070	0.0056	0.0015	−0.0043	0.0006
0.4894	0.0050	−0.0008	−0.0276	−0.0125	0.0193	0.0097	−0.0099	−0.0104	0.0026	0.0077
0.5665	0.0010	0.0035	−0.0034	−0.0329	−0.0107	0.0162	0.0137	0.0009	−0.0058	−0.0071
0.6515	−0.0020	0.0016	0.0269	−0.0224	−0.0163	−0.0079	−0.0038	0.0072	0.0067	0.0043
0.7459	−0.0031	−0.0013	0.0381	0.0108	0.0037	−0.0160	−0.0105	−0.0070	−0.0065	−0.0007
0.8511	0.0088	−0.0066	0.0238	0.0484	0.0299	0.0095	0.0211	0.0023	0.0126	−0.0044
0.9683	0.0025	0.0040	−0.0653	0.0189	−0.0095	0.0073	−0.0134	0.0066	−0.0223	−0.0142
1.0984	−0.0096	−0.0015	−0.0223	−0.0425	−0.0325	−0.0407	0.0520	−0.0127	−0.0220	0.0408
1.3980	−0.0043	0.0033	0.0048	0.0346	−0.0497	−0.0195	−0.0134	0.0265	0.0310	0.0155
1.5668	−0.0041	−0.0140	−0.0197	0.0303	0.0459	0.0043	0.0232	0.0050	0.0199	0.0026
1.7476	0.0069	0.0205	−0.0447	−0.0583	−0.0428	0.0014	0.0092	0.0090	0.0054	−0.0011
1.9397	−0.0132	−0.0358	0.0521	−0.0315	0.0380	−0.0102	−0.0003	0.0157	−0.0074	−0.0014
2.1428	−0.0384	0.0204	−0.0797	0.0922	−0.0619	−0.0546	0.0604	−0.0223	0.0050	0.0217
2.3564	0.0235	0.0179	−0.0055	−0.0941	0.0119	0.0728	0.0453	−0.0255	−0.0161	−0.0010
2.8148	−0.0074	0.0071	−0.0370	0.0074	−0.0170	0.0323	0.0005	−0.0024	0.0123	−0.0109

Appendix D

The equilibrium of the forces acting on the each floater of the array leads to the following system of differential equations of motion in the frequency domain, ($q = 1, 2, \dots, N$), i.e.,:

$$\sum_{p=1}^{N}\sum_{j=1}^{6}\left(\delta_{p,q}m_{kj}^{q}+a_{k,j}^{q,p}\right)\ddot{x}_{j0}^{p}+\left(\beta_{k,j}^{q,p}+\delta_{p,q}\delta_{j,3}b_{PTO}^{p}\right)\dot{x}_{j0}^{p}+\delta_{p,q}c_{kj}^{q}x_{j0}^{p}=F_{D,k}^{q}, \; k=1,\ldots,6 \qquad (A1)$$

where x_{j0}^{p} is the 6–degree displacement vector of the p floater of the array; m_{kj}^{q} is the mass matrix of the q floater; $a_{k,j}^{q,p}$ is the frequency–dependent hydrodynamic mass matrix and $\beta_{k,j}^{q,p}$ is the frequency–dependent damping matrix of the q floater in the kth direction due to the forced oscillation of the p floater in the jth direction (see Equation (19)); c_{kj}^{q} is the stiffness matrix; $F_{D,k}^{q}$ represents the exciting force on the q floater in the kth direction (see Equation (17)); b_{PTO}^{p} is the damping coefficients that originate from the PTO mechanism and $\delta_{p,q}$, $\delta_{j,3}$ are Kronecker delta.

References

1. McCormick, M.E. *Ocean Wave Energy Conversion*; Courier Corporation: North Chelmsford, MA, USA, 1981.
2. Pelc, R.; Fujita, R.M. Renewable energy from the ocean. *Mar. Policy* **2002**, *26*, 471–479. [CrossRef]
3. Falnes, J. A review of wave-energy extraction. *Mar. Struct.* **2007**, *20*, 185–201. [CrossRef]
4. Falcao, A.F.O. Wave energy utilization: A review of the technologies. *Renew. Sustain. Energy Rev.* **2010**, *14*, 899–918. [CrossRef]
5. Aderinto, T.; Li, H. Ocean wave energy converters: Status and Challenges. *Energies* **2008**, *11*, 1250. [CrossRef]
6. Giebhardt, J.; Kracht, P.; Dick, C.; Salcedo, F. *Report on Grid Integration and Power Quality Testing*; Deliverable 4.3 Final; Marinet: Singapore, 2014.
7. Magagna, D.; Uihlein, A. Ocean energy development in Europe: Current status and future perspectives. *Int. J. Mar. Energy* **2015**, *11*, 84–104. [CrossRef]
8. Mustapa, M.A.; Yaakob, O.B.; Ahmed, Y.M.M.; Rheem, C.K.; Koh, K.K.; Faizul, A.A. Wave energy device and breakwater integration: A review. *Renew. Sustain. Energy Rev.* **2017**, *77*, 43–58. [CrossRef]
9. Rusu, E. Evaluation of the wave energy conversion in various coastal environments. *Energies* **2014**, *7*, 4002–4018. [CrossRef]
10. Cascajo, R.; Garcia, E.; Quiles, E.; Correcher, A.; Morant, F. Integration of marine wave energy convertres into seaports: A case study in port of Valencia. *Energies* **2019**, *12*, 787. [CrossRef]
11. Mavrakos, S.A.; Katsaounis, G.M.; Nielsen, K.; Lemonis, G. Numerical performance investigation of an array of heaving wave power converters in front of a vertical breakwater. In Proceedings of the 14th International Offshore and Polar Engineering Conference (ISOPE 2004), Toulon, France, 23–28 May 2004.
12. Mavrakos, S.A.; Katsaounis, G.M.; Kladas, A.; Kimoulakis, N. Numerical and experimental investigation of performance of heaving WECs coupled with DC generators. In Proceedings of the 9th European Wave and Tidal Energy Conference, Southampton, UK, 5–9 September 2011.
13. Teng, B.; Ning, D.Z. Wave diffraction from a uniform cylinder in front of a vertical wall. *Ocean Eng.* **2003**, *21*, 48–52.
14. Teng, B.; Ning, D.Z.; Zhang, X.T. Wave radiation by a uniform cylinder in front of a vertical wall. *Ocean Eng.* **2004**, *31*, 201–224. [CrossRef]
15. Zheng, S.; Zhang, Y. Wave diffraction from a truncated cylinder in front of a vertical wall. *Ocean Eng.* **2015**, *104*, 329–343. [CrossRef]
16. Zheng, S.; Zhang, Y. Wave radiation from a truncated cylinder in front of a vertical wall. *Ocean Eng.* **2016**, *111*, 602–614. [CrossRef]
17. Schay, J.; Bhattacharjee, J.; Soares, C. Numerical modelling of a heaving point absorber in front of a vertical wall. In Proceedings of the 32nd International Conference on Ocean, Offshore and Artic Engineering (OMAE 2013), Nantes, France, 9–14 June 2013. [CrossRef]
18. He, F.; Huang, Z. Using an oscillating water column structure to reduce wave reflection from vertical wall. *J. Waterw. Port Coast. Ocean Eng.* **2016**, *142*. [CrossRef]
19. Martins-rivas, H.; Mei, C.C. Wave power extraction from an oscillating water column at the tip of a breakwater. *J. Fluid Mech.* **2009**, *626*, 395–414. [CrossRef]
20. Martins-rivas, H.; Mei, C.C. Wave power extraction from an oscillating water column along a straight coast. *Ocean Eng.* **2009**, *36*, 426–433. [CrossRef]
21. Howe, D.; Nader, J.R. OWC WEC integrated within a breakwater versus isolated: Experimental and numerical theoretical study. *Mar. Energy* **2017**, *20*, 165–182. [CrossRef]

22. Naty, S.; Viviano, A.; Foti, E. Wave energy exploitation system integrated in the coastal structure of a Mediterranean port. *Sustainability* **2016**, *8*, 1342. [CrossRef]
23. Naty, S.; Viviano, A.; Foti, E. Feasibility study of a WEC integrated in the port of Giardini Naxos, Italy. In Proceedings of the 35th Conference on Coastal Engineering, Antalya, Turkey, 17–20 November 2016.
24. He, F.; Huang, Z.; Law, A.W.K. Hydrodynamic performance of a rectangular floating breakwater with and without pneumatic chambers: An experimental study. *Ocean Eng.* **2012**, *51*, 16–27. [CrossRef]
25. Zheng, X.; Zeng, Q.; Liu, Z. Hydrodynamic performance of rectangular heaving buoys for an integrated floating breakwater. *J. Mar. Sci. Eng.* **2019**, *7*, 239. [CrossRef]
26. He, F.; Zhang, H.; Zhao, J.; Zheng, S.; Iglesias, G. Hydrodynamic performance of a pile-supported OWC breakwater: An analytical study. *Appl. Ocean Res.* **2019**, *88*, 326–340. [CrossRef]
27. Zhao, X.L.; Ning, D.Z.; Zhang, C.W.; Liu, Y.Y.; Kang, H.G. Analytical study on an oscillating buoy wave energy converter integrated into a fixed box-type breakwater. *Math. Probl. Eng.* **2017**, *2017*, 1–9. [CrossRef]
28. Zhao, X.L.; Ning, D.Z.; Liang, D.F. Experimental investigation on hydrodynamic performance of a breakwater-integrated WEC system. *Ocean Eng.* **2019**, *171*, 25–32. [CrossRef]
29. Salter, S.; Taylor, J.; Caldwell, N. Power conversion mechanisms for wave energy. *J. Eng. Marit. Environ.* **2002**, *216*, 1–27. [CrossRef]
30. Falnes, J.; Hals, J. Heaving buoys, point absorbers and arrays. *Philos. Trans. R. Soc. A* **2012**, *370*, 246. [CrossRef] [PubMed]
31. Rahmati, M.T.; Aggidis, G.A. Numerical and experimental analysis of the power output of a point absorber wave energy converter in irregular waves. *Ocean Eng.* **2016**, *111*, 483–492. [CrossRef]
32. Tampier, G.; Grueter, L. Hydrodynamic analysis of a heaving wave energy converter. *Int. J. Mar. Energy* **2017**, *19*, 304–318. [CrossRef]
33. Son, D.; Yeung, R. Real time implementation and validation of optimal damping control for a permanent magnet linear generator in wave energy extraction. *Appl. Energy* **2017**, *208*, 571–579. [CrossRef]
34. Gaspar, J.; Calvario, M.; Kamarlouei, M.; Soares, C. Power take-off concept for wave energy converters based on oil-hydraulic transformer units. *Renew. Energy* **2016**, *86*, 1232–1246. [CrossRef]
35. Spring, B.W.; Monkmeyer, P.L. Interaction of plane waves with a row of cylinders. In Proceedings of the 3rd Specialty Conference of Civil Engineering in Oceans ASCE, Newark, DE, USA, 9–12 June 1975; pp. 979–998.
36. Yeung, R.W.; Sphaier, S.H. Wave-interference effects on a truncated cylinder in a channel. *J. Eng. Math.* **1989**, *23*, 95–117. [CrossRef]
37. Yeung, R.W.; Sphaier, S.H. Wave-interference effects on a floating body in a towing tank. In Proceedings of the 4th International Symposium on Practical Design of Ships and Mobile Units (PRADS), Varna, Bulgaria, 23–28 October 1989.
38. Mavrakos, S.A. The scattered wave field by vertical cylinders in a narrow tank. In Proceedings of the 4th National Symposium on Theoretical and Applied Mechanics, Xanthi, Greece, 26–29 June 1995; Volume II, pp. 819–829.
39. Twersky, V. Multiple scattering of radiation by an arbitrary configuration of parallel cylinders. *J. Acoust. Soc. Am.* **1952**, *24*, 42. [CrossRef]
40. Okhusu, M. Hydrodynamic forces on multiple cylinders in waves. In Proceedings of the International Symposium on the Dynamics of Marine Vehicles and Structures in Waves, London, UK, 1–5 April 1974.
41. Mavrakos, S.A.; Koumoutsakos, P. Hydrodynamic interaction among vertical axisymmetric bodies restrained in waves. *Appl. Ocean Res.* **1987**, *9*, 128–140. [CrossRef]
42. Mavrakos, S.A. Hydrodynamic coefficients for groups of interacting vertical axisymmetric bodies. *Ocean Eng.* **1991**, *18*, 485–515. [CrossRef]
43. Mavrakos, S.A.; McIver, P. Comparison of methods for computing hydrodynamic characteristics of arrays of wave power devices. *Appl. Ocean Res.* **1997**, *19*, 283–291. [CrossRef]
44. Konispoliatis, D.N.; Mavrakos, S.A. Hydrodynamic analysis of an array of interacting free–floating oscillating water column devices. *Ocean Eng.* **2016**, *111*, 179–197. [CrossRef]
45. Kokkinowrachos, K.; Mavrakos, S.; Asorakos, S. Behavior of vertical bodies of revolution in waves. *Ocean Eng.* **1986**, *13*, 505–538. [CrossRef]
46. Newman, J.N. *Marine Hydrodynamics*; MIT Press: Cambridge, UK, 1977.
47. Faltinsen, O.M. *Sea Loads on Ships and Offshore Structures*; Cambridge University Press: Cambridge, UK, 1990.

48. Loukogeorgaki, E.; Chatjigeorgiou, I. Hydrodynamic performance of an array of wave energy converters in front of a vertical wall. In Proceedings of the 13th European Wave and Tidal Energy Conference (EWTEC), Naples, Italy, 1–6 September 2019.
49. Nader, J.R.; Zhu, S.P.; Cooper, P.; Steppenbelt, B. A finite element study of the efficiency of arrays of oscillating water column wave energy converters. *Ocean Eng.* **2012**, *43*, 72–81. [CrossRef]
50. Falnes, J. *Ocean. Waves and Oscillating Systems: Linear Interactions Including Wave-Energy Extraction*; Cambridge University Press: Cambridge, UK, 2002.

Journal of
*Marine Science
and Engineering*

Article

Wave Energy Converter Power Take-Off System Scaling and Physical Modelling

Gianmaria Giannini [1,2,*], Irina Temiz [3], Paulo Rosa-Santos [1,2], Zahra Shahroozi [3],
Victor Ramos [1,2], Malin Göteman [3], Jens Engström [3], Sandy Day [4] and Francisco Taveira-Pinto [1,2]

[1]   Department of Civil Engineering, Faculty of Engineering of the University of Porto (FEUP),
      4200-465 Porto, Portugal; pjrsantos@fe.up.pt (P.R.-S.); jvrc@fe.up.pt (V.R.); fpinto@fe.up.pt (F.T.-P.)
[2]   Interdisciplinary Centre of Marine and Environmental Research of the University of Porto (CIIMAR),
      4200-465 Porto, Portugal
[3]   Department of Electrical Engineering, Uppsala University, P.O. Box 256, SE-751 05 Uppsala, Sweden;
      Irina.Temiz@angstrom.uu.se (I.T.); zahra.shahroozi@angstrom.uu.se (Z.S.);
      malin.goteman@angstrom.uu.se (M.G.); jens.engstrom@angstrom.uu.se (J.E.)
[4]   Department of Naval Architecture, Ocean and Marine Engineering, University of Strathclyde,
      Glasgow G4 0LZ, UK; sandy.day@strath.ac.uk
*   Correspondence: gianmaria@fe.up.pt

Received: 28 July 2020; Accepted: 13 August 2020; Published: 20 August 2020

Abstract: Absorbing wave power from oceans for producing a usable form of energy represents an attractive challenge, which for the most part concerns the development and integration, in a wave energy device, of a reliable, efficient and cost-effective power take-off mechanism. During the various stages of progress, for assessing a wave energy device, it is convenient to carry out experimental testing that, opportunely, takes into account the realistic behaviour of the power take-off mechanism at a small scale. To successfully replicate and assess the power take-off, good practices need to be implemented aiming to correctly scale and evaluate the power take-off mechanism and its behaviour. The present paper aims to explore and propose solutions that can be applied for reproducing and assessing the power take-off element during experimental studies, namely experimental set-ups enhancements, calibration practices, and error estimation methods. A series of recommendations on how to practically organize and carry out experiments were identified and three case studies are briefly covered. It was found that, despite specific options that can be strictly technology-dependent, various recommendations could be universally applicable.

Keywords: power take-off damping; wave power device; experimental testing; PTO simulator; uncertainty analysis; wave energy testing; experimental set-up; calibration

---

1. Introduction

Harvesting power from ocean waves represents a fascinating challenge. Over the years, many wave energy converter (WEC) concepts have been proposed [1,2]. Figure 1 summarizes different examples of the most successful WECs projects. WECs can be categorized depending on how the power take-off (PTO) system is activated. In general, a PTO requires a counter-reaction force to work; thus, two main groups of devices can be defined: Earth-reacting and self-reacting. Moreover, devices can be subdivided in fixed structures, floating, and submerged. For all WECs, the PTO system is the most important component, which needs to be developed as an integral part. The PTO influences the dynamics of a WEC and its reliability, performance and cost are critical factors. Consequently, early experimental testing of the PTO at laboratory scale becomes essential for validation of WEC development. Therefore, it is very important to consolidate guidelines and theory for WEC scaling [3] with particular attention to the PTO system.

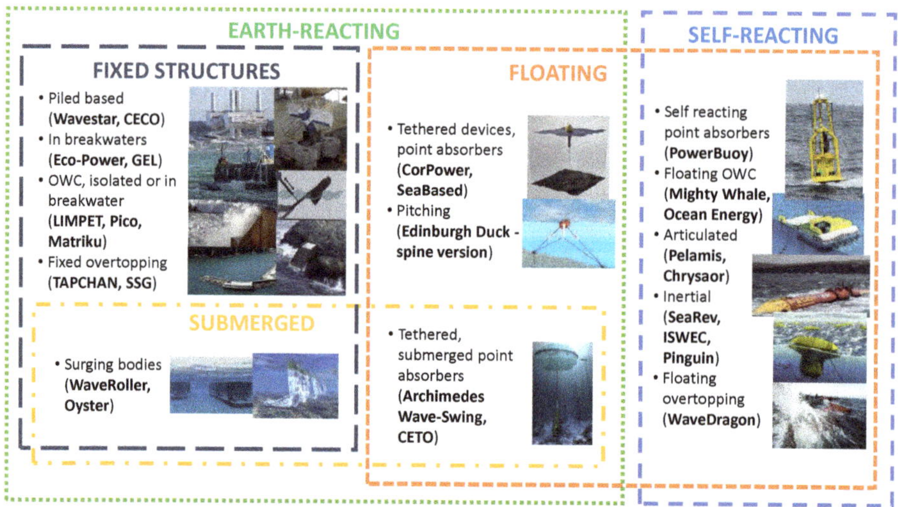

Figure 1. Types of wave energy converters and examples.

Reproducing the effect of the PTO system at laboratory scale is a challenging task. A first idea would be related to using a complex miniaturized PTO mechanism, which could realistically produce little amounts of electricity. Optionally, PTO systems are often physically implemented as an active damper element, which simulates the PTO reaction forces (PTO simulator) [4]. In all cases, it is worth noting that the results obtained could present high uncertainties, owing to different factors such as unpredictable friction losses between moving parts, signal noise, instrument resolution, accuracy and repeatability. In general, the PTO is monitored by a set of sensors and the absorbed power is assessed by multiple measurements. In this situation, uncertainties from multiple sensors have to be combined and the global uncertainty related to the power estimation further amplifies.

To limit and correctly quantify uncertainties related to model-scale PTOs, three types of actions can be implemented. The firsts, concern following good practices and recommendations on how to construct laboratory-scale PTO systems. The second set of actions relates to essential calibration procedures needed at the beginning of experimental tests. In this sense, correct calibration and characterization of the PTO damping with early tests becomes indispensable before moving to the actual tests in water. Finally, the third type of action concerns a correct quantification of uncertainties following common practice and known theory.

In particular, regarding the PTO forces and damping values, limited literature deals with the procedures to assess uncertainties related to empirical quantities. Existing documentation mainly focuses on wave tank testing of conventional floating structures or ships models, e.g., International Towing Tank Conference (ITTC) guidelines [5,6]. Nevertheless, requirements of wave energy testing are dissimilar from conventional structures [3]. For the case of marine renewable energy devices, introductory guidance is provided by [7–9]. Similarly, few first general standards for experimental testing of marine renewable energy systems are available, e.g., European Marine Energy Centre [10]. Nevertheless, all published guidance mostly applies to real scale experimental tests and only little focuses on the specific problem of scaling PTO components for laboratory testing. Moreover, many PTO control methodologies have been proposed, but experiments at laboratory dealing with PTO control, due to scaling reasons, are rather difficult to be performed accurately [4].

Against this backdrop, the present paper focuses on reviewing basic theory and good scientific practices for reproducing PTO systems at the laboratory (Sections 2 and 3). Three case studies on direct-drive electrical and mechanical types of PTO are also analysed, to provide examples of possible experimental arrangements for laboratory-scale PTO testing (Section 4). Finally, brief concluding remarks based on learnings are provided (Section 5).

2. State-Of-The-Art

2.1. Types of PTO Systems

Different types of PTOs exist, and depending on the category, these can be easier to scale, more difficult, or at cases, may not be scalable. The most common PTOs are based on air turbines, hydraulic, electrical direct-drive, and mechanical systems.

Air turbines systems are implemented in WECs operating under the principle of oscillating water columns (OWCs). This type of PTO is often tested with 1:5 to 1:9 model scales [11]. Dealing with OWC PTOs at smaller scales is a problematic task since it must take into account complex wave-structure interaction phenomena, such as air compressibility and, fluid and PTO dynamics [3,12]. Therefore, different approaches have been adopted, either by modifying proportions of the OWC chamber with respect to the real scale or by keeping the same ratio between width, height and length [11]. As an alternative, deformable air chambers can be used [13]. In most cases, the PTO is simulated at model scale by using an orifice or porous media [14]. The pressure drop due to the orifice is monitored by internal and external pressure sensors. Due to scaling reasons, as explained in detail in [12], external air pressure at model scale should be increased for allowing physical similarities. This aspect may be resolved by connecting an external air reservoir to the OWC. For fixed OWC models, this solution could be easy to implement. However, given the dimensions and weight of the air reservoir required, the solution may not be simple to implement in floating OWCs. Assuming that the model set-up correctly represents the real scale OWC device ("physically"), the power from the simulated orifice PTO can be estimated from measuring the pressure drop (Δp), according to the following expression:

$$P_{pto} = \Delta p\, Q \tag{1}$$

where Q is the volumetric flow rate of the air passing through the orifice. An orifice-based PTO simulator can be calibrated with preliminary regular wave tests in water through measuring the internal free surface elevation with wave probes [15]. In particular, measuring Q is difficult because flow meters introduce head losses that are of the same magnitude of PTO damping. As a solution, Q can be characterized preliminarily as a function of the pressure drop. Thus, Q became a known entity in Equation (1).

Hydraulic systems, in general, are only scalable until a certain extent. Small scale hydraulic PTOs are unfeasible due to the requirement of impractical high fluid pressure. Concerning these systems, there are a series of extra issues to be considered, for instance, friction drag of seals and viscous drag due to small holes. These issues are very significant and may radically affect PTO efficiency [16].

Conversely, direct-drive electrical and mechanical types of PTO are normally better scalable. These types of systems are applied to WECs that have moving components, normally, of relatively large dimensions and mass. In this situation, Froude similitudes are typically readily applicable. However, depending on the complexity of the PTO, the mechanical friction (dynamic and static) and the inertial effects, the scaling of such PTOs could still be a laborious process. Overall, direct-drive electrical and mechanical PTOs are among the most widely used type and can be better scaled than other PTO systems. As such PTOs are implemented in the later discussed case studies, the rest of the paper mainly focuses on these.

2.2. Scaling Laws

Small-scale models of WECs should represent reliably real scale physics. To achieve so a series of approaches can be followed. In theory, a previous dimensional analysis should be carried out to design the scaled model for ensuring similarities. In practice, for PTO modelling dimensional analysis can be rather complex and unsuitable to be implemented. Thus, other solutions for designing the model can be implemented, for instance involving the adoption of assumptions and the application of specific direct scaling laws. To achieve consistency between the laboratory model and its real scale version geometrical, kinematic, and dynamic similitudes should be fulfilled. Focusing on offshore structure testing it may be useful to calculate, for instance, the dimensionless numbers referred as Froude (Fr), Reynolds (Re), Mach's (Mn), Weber's (Wn), Keulegan-Carpenter (KC) and Strouhall (St), which represent force ratios of inertia/gravity (Fr), inertia/viscous (Re), inertia/elasticity (Mn), inertia/surface tension (Wn), drag/inertia (KC) and the non-dimensional vortex shedding frequency (St), respectively. The first two dimensionless numbers are the most common and relevant for the PTO of the case studies later described thus only these are covered with some details next.

Froude and Reynolds numbers relative to the real scale prototype and its model scaled version should be kept as much as constant as possible. Keeping both these numbers constant at the same time, is not possible because it would imply the use of a fluid that does not exist in reality, solutions may not be suitable for practical experimental testing or are excessively expensive. Thus, for practically scaling a WEC, a particular compromise should be established. The Froude number, *Fr* and Reynolds number, *Re*, are defined as:

$$Fr = \frac{U}{\sqrt{gL}} \tag{2}$$

$$Re = \frac{\rho U L}{\mu} \tag{3}$$

where U represents the characteristic velocity of the fluid, L the characteristic length of the device, g the acceleration of gravity, ρ the fluid density and μ the dynamic viscosity of the fluid. The Froude number (Fr) indicates the importance of inertial forces relative to gravity forces. The Reynolds number (Re), instead, provides a measure of the importance of inertial forces relative to the viscous forces. Depending on the scope of the experimental work, either the Froude number or the Reynolds number is normally kept analogous to the real scale case.

For wet WEC testing, if a suitable scale is selected, the gravity forces normally are significantly higher compared to viscous forces. The Reynold number can indicate the validity of Froude scaling. As a rough indication, a Reynolds number minimum of 10^5, for the real scale case, indicates good applicability of Froude scaling laws [3]. In such circumstance, for experimental studies on WECs, it is commonly accepted to use primarily Froude scaling laws, which are reported in Table 1. In this table, s represents the geometrical scaling factor. For Re lower than 10^5, viscous forces may be relevant and Froude scaling laws may not be directly applicable. In this case, viscous forces may be significant and need to be taken somehow into account, for instance with the aid of a numerical model as for [17]. To note that for comparing model scale results obtained at a hydrodynamics laboratory with ocean prototype results, it is required to consider as well a correction factor ($r = \rho_{ocean}/\rho_{tank}$) that takes into account the density difference between saltwater of the ocean and freshwater of the wave tank [18].

As can be observed in Table 1, the power scales by the $s^{3.5}$ law, meaning that for small model scales the multiplication factor (needed for converting laboratory-scale values into real scale quantities) can be very large, as shown in Figure 2. In consequence, the power produced by a WEC model (at small scale) is a significantly small quantity and uncertainties related to the simulation of the PTO may highly distort the expected PTO behaviour.

Table 1. Common scaling factors used in WEC experimental testing campaigns.

Quantity	Scaling Law
Linear displacement	s
Angular displacement	1
Translational velocity	$s^{0.5}$
Angular velocity	$s^{-0.5}$
Translational acceleration	1
Angular acceleration	s^{-1}
Mass	s^3
Force	s^3
Torque	s^4
Power	$s^{3.5}$
Linear stiffness	s^2
Angular stiffness	s^4
Linear damping	$s^{2.5}$
Angular damping	$s^{4.5}$
Wave height and length	s
Wave period	$s^{0.5}$
Wave frequency	$s^{-0.5}$
Power density	$s^{2.5}$

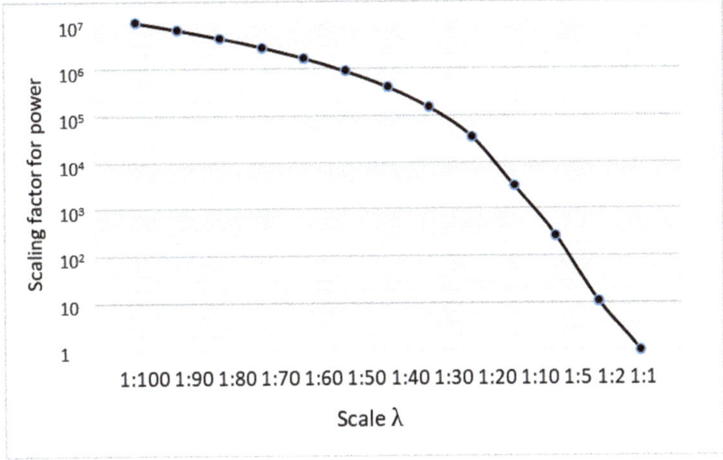

Figure 2. Froude scaling factor for power (note: vertical axis has a logarithmic scale).

2.3. Experimental Scale Selection

In general, before obtaining the final design of a commercial WEC, extensive research must be carried out by using both numerical and experimental testing. Research and development results need to be regularly validated with data from experimental campaigns performed at different model scales due to budget reasons. Thus, experimental work is crucial for ensuring the validity of calculations, concepts proposed and make the proof of system functionalities. As proposed by the European Marine Energy Centre, at least 5 development phases are required [10], Table 2. During these phases, depending on the scope and economic resources available, a different type of PTO system can be used.

Table 2. Phases for WEC development, adapted from the European Marine Energy Centre [10].

	Phase 1 Validation Model			Phase 2 Design Model	Phase 3 Process Model	Phase 4 Prototype	Phase 5 Full Size
Scale (λ)	1:25–100			1:10–25	1:3–10	1:2	1:1
Technology readiness level (TRL)	1–3			4–5		6–7	8–9
Testing environment	2D flume and 3D wave tanks			3D basin	Sheltered sea site (benign)	Exposed sea site	Open sea location
Duration of tests including analysis	1–3 weeks	1–3 months	1–3 months	6–12 months	6–18 months	12–36 months	1–5 years
Typical no. of tests	50–500	250–500	100–250	100–250	50–250	continuous	statistical sample
Indicative budget (€,000)	1–5	25–75	25–50	50–250	1000–2500	5000–10,000	2500–7500
Conditions to test	Regular waves Up to 5 irregular sea states tests (unidirectional)	Irregular sea states (short and log crested, multidirectional sea states)		Pilot site sea spectra Long and short crested classical seas (multidirectional sea states)	Extended test at sea to ensure all seaways are included	Full evaluation	Full evaluation
PTO system	PTO simulator			Miniaturized PTO		Real PTO	Certified PTO

At phase 1, a geometrical scale ($\lambda = 1/s$) within the range of $\lambda = 1 : 25 - 100$ can be adopted, e.g., [19–21]. At this stage, the main aim of the experimental work normally concerns a proof-of-concept, initial preliminary assessments on performances and numerical model validation. For such scales, rarely, a realistic PTO is implemented. In this case, a PTO simulator is normally used. This PTO simulator can be, for instance, an electric motor working as an active damper with a feedback control loop, mechanical breaks or hydraulic dampers.

For phases 2 and 3, model scales as $\lambda = 1 : 3 - 25$ can be used, e.g., [22–24]. In this case, the PTO can be either, a simulator or, as well, eventually a fully functional scaled generator. However, at this stage, the best scale to use is highly dependent on the type of technology proposed and technology readiness level (TRL), which for marine renewable devices is defined by [25]. By increasing the physical model size, the PTO scaling effects are progressively less relevant. For scales close to (and eventually less than) $\lambda = 1 : 25$, a miniaturized PTO could be representative of the real scale counterpart, but only if high-quality industrial-grade experimental equipment is used. At phases 2 and 3, the realistic behaviour of the PTO system needs to be validated. Thus, the PTO electronics could be tested, eventually also implementing possible control methodologies. In this way, expected performances can be proved, in particular, using irregular sea states. In this occasion, it may be worth also testing the PTO survival control mode during a set of extreme sea states. Phases 4 and 5 involve prototype testing and, hence, are not very related to the scope of this paper.

3. Common Practices

3.1. Experimental Set-Ups

In particular, at small scales, friction losses within the PTO system can be the main source of uncertainties. Friction effects within moving components are neither easily scalable nor linear. Thus, unwanted friction in all cases needs to be reduced for the most. Each component of the experimental set-up needs to be selected by keeping in mind to reduce friction losses. Within the entire experimental rig, friction losses, if substantial, can significantly modify the motion of the WEC even more than the expected PTO damping effect. To reduce friction between moving components, it is advisable to use very low friction bearings such as hydrostatic [8] or ceramic bearings [26]. Besides, choosing bearings of larger radius could allow a further reduction of friction losses.

Different options for the reproduction of the PTO system on a laboratory scale exist. Apart from using miniaturized PTO systems, which can be feasible at larger experimental scales (greater than 1:25), PTO simulators may be a feasible solution. The lasts allow for reproducing a realistic PTO behaviour, at least from the theoretical standpoint, skipping errors relative to friction losses. In this case, representative PTO forces are targeted. A review of the topic is provided by Beatty et al. [4]. There are various advantages by physically model PTOs at laboratory scales by using active damper systems. The implementation of such PTO simulators allows for experimental flexibility. By using such options, the PTO force can be highly customized to specific needs. For instance, an active PTO simulator can be an electric motor working as an active damper with a feedback control loop [8,26] or electromagnetic type of system based on eddy currents [27,28]. While the last kind of system requires advanced electromechanical design, PTO systems based on the use of electric motors, instead, can be set up by using off-the-shelf components, for example as implemented by Zurkinden et al. [22], who used a feedback control loop and a PID set-up (close-control loop) for emulating the PTO force, e.g., as for Figure 3. To well represent the PTO force using a control loop approach, very low-latency control electronics are required. Dedicated computer resources for the PTO control should be used, thereby; latencies can be minimized, i.e., within orders of few milliseconds. A controller, connected as for Figure 3, corrects the PTO force by processing information (rotation speed and direction) provided by the encoder or tachometer element. Moreover, through a look-up function, it is possible to implement specific adjustments depending on previous calibration and characterization of the electric motor. Active PTO simulators, in theory, could allow testing at model scale any type of WEC's PTO control strategy, and eventually, almost reproducing faithfully the behaviour of the real scale prototype.

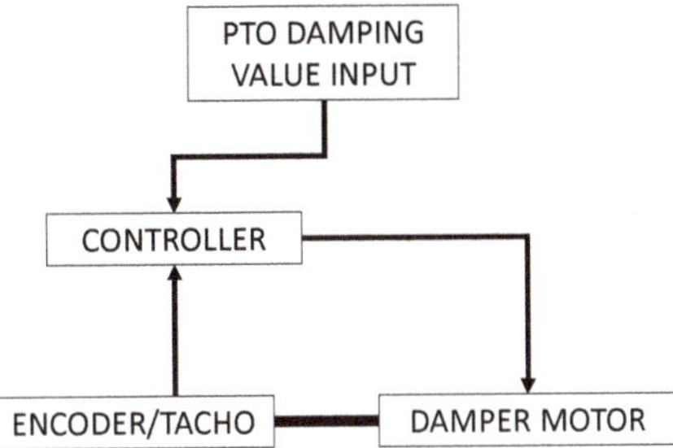

Figure 3. Scheme of connections for closed-loop control of an electric motor PTO simulator.

Instead of using electronic-based arrangements, 'pure' mechanical systems can be used, for example as the one implemented by Troch et al. [29], where PTO simulators are mechanical brakes. This approach can be referred to as passive PTO simulator [4]. A pure mechanical option can be an attracting and easy solution to implement, during the early stages of WEC development. However, the use of mechanical dampers normally implies a nonlinear PTO force (F_{PTO}), which can be represented by means of Coulomb damping. For a heaving point absorber this can be found by [30]:

$$F_{PTO} = -\mu F_N sign(\dot{z}) \tag{4}$$

where μ is the coefficient of the mechanical friction used, F_N the normal force that is given by the mechanical brake and $sign(\dot{z})$ the sign of the heave velocity of the floater.

In general, when constructing a PTO experimental set-up, the following recommendations can be provided:

- Minimize unwanted friction;
- Minimize inertia of PTO components;
- Minimize tolerances between components;
- Use high quality industrial experimental equipment;
- Ensure rigidity of fixed components, unless flexibility/deformation is assessed;
- Use parts that are machine built and, eventually, are made of mechanically advantageous materials such as carbon fibre, aluminium or stainless steel;
- Reduce complexities.

3.2. Calibration Procedures

Once a PTO set-up is defined, a precise PTO calibration methodology is required to regulate and analyse the PTO force. Through calibration, a force magnitude close to the target PTO force can be obtained. Besides, the PTO force can be as well characterized, determining its relationship with the displacement and velocity of the PTO driving elements. For this purpose, different calibration methodologies can be implemented.

Normally, calibration is performed by running a series of dry-tests, which do not involve water and waves. If possible, it would be best to calibrate and assess the PTO behaviour, before adding further complexities, related to the full waves-structure interaction problem. Besides, at this stage, the sensors to be used for monitoring the PTO system must be accurately configured and calibrated. All PTO control units and sensors should be preliminarily tested, preferably taking into account circumstances, as for actual tests, in water, e.g., data acquisition equipment and loggers connections, which will be later used and may determine signal noise. It is advisable to set up a PTO simulator rig where all sensors are installed. The rig can be extensively calibrated during dry-tests and successively transferred, as it is, to the wave tank for later tests, with the entire WEC in water. Following this approach, later modifications can be reduced and, therefore, the calibration of the PTO and relative sensors can be preserved. Later calibration checks are always beneficial, but these not always can be done once the WEC model is in water.

PTOs for oscillating WECs, initially, can be calibrated with drop-weights tests. For instance, if the PTO simulator is an electric motor, a worm drive can be installed on the motor's shaft. Weights attached to a wire can be used to drive the motor used as a damper. Given a specific weight value, and the falling speed, the PTO damping constant relative to a specific setting (current and voltage) can be inferred. The PTO force in this way can be tuned for several values within the desired range.

However, by only using drop tests, the effect of varying motor direction of rotation and its acceleration may not be sufficient. For this reason, it may be opportune to perform further PTO tuning by oscillations type of tests. For this purpose, spring elements can be temporarily added to the PTO rig set-up. Oscillations at target frequencies can be achieved by choosing a combination of correct weights and springs having suitable stiffness. Besides, to understand better uncertainties related to the PTO, it is of fundamental importance to run calibration tests several times. The more repetitions are conducted, the better level of confidence can be assumed for later calculations of the expanded uncertainty value. In later case studies section (Section 4), examples of PTO physical models set-up and calibration methodologies are presented.

In general, for calibration of PTO systems, it can be recommended to:

- Explore target PTO force values and velocities ranges;
- Apart from linear velocity, PTO's oscillation motion should be assessed;
- Perform as many as feasible repetitions;
- Test, disassemble, reassemble and re-test;
- Keep the PTO rig unchanged when it is needed to be transferred from a dry-test facility to the wave tank;
- Possibly, daily re-check of sensors' calibration (during actual tests in water).

3.3. Experimental Errors Evaluation

Experimental error evaluation practices can be divided into informal and formal. In informal approaches, the assessment of the accuracy of experimental tests can be limited to the description of experiments, mainly concerning the qualitative comparison of results. In this case, experimental errors can be roughly estimated by expert judgement without following a specific standard method. An expert opinion/evaluation may be based on previous experience or observations. Such informal approaches may be reported by a description of the methodology implemented and may only apply to specific experiments for which no standard error estimation procedures are implemented or can be identified (e.g., new PTO concepts). This kind of approach can be adopted for initial research stages, for instance, when proof-of-concept is investigated and only coarse calculations are undertaken.

As informal approaches for estimating errors could be very dependent on individual awareness, these may not be always well accepted at more advanced stages of development. Formal uncertainty analysis methods should instead be preferred to informal approaches. It could be good practice to include a full uncertainty analysis section within a report or publication for supporting experimental results. Confidence intervals relative to direct and indirect, empirical measurements can be reported within independent tables and/or as error bars inside graphs of experimental results.

Practical guidance on formal uncertainty analysis that can be applied during experimental works on WECs is provided by ITTC [5,9] and EquiMar [31]. For what concerns direct measurements related to the PTO system, the standard uncertainty u_s should be evaluated using:

$$u_s = \sqrt{(u_{s-A})^2 + (u_{s-B})^2}$$ (5)

where u_{s-A} and u_{s-B} are the Type A and Type B uncertainties, respectively. u_{s-A} reflects the repeatability of the experiment and statistical errors, and can be calculated as follows:

$$u_{s-A} = \frac{s}{\sqrt{n}}$$ (6)

where n is the number of repetitions of tests and s is the standard deviation, which may be defined as:

$$s = \sqrt{\frac{\sum_{k=1}^{n}(q_k - \bar{q})}{n-1}}$$ (7)

where q_k is the empirical measurement associated with a specific k test and $\bar{q}$ is the average value of all taken measurements. The quantity $(q_k - \bar{q})$ is the relative error of a specific sample.

Differently, the uncertainty Type B (u_{s-B}) is estimated by prior experience, calibration of equipment, manufacturers' specifications and other relevant information.

Methods based on regression analysis (e.g., linear regression), as mentioned in [5], can be used to estimate u_{s-B}. These methods concern about fitting a known curve into calibration data to calculate residuals. For linear regression analysis, residuals R_i can be calculated as:

$$R_i = y_i - a - bx_i$$ (8)

which represents a difference between the measured values y_i and a straight line $y = (a + bx_i)$.

The uncertainty Type B value (u_{s-B}) can then be calculated as:

$$u_{s-B} = \sqrt{\frac{SS_R}{n-2}}$$ (9)

where SS_R is the sum of residuals (R_i) squared.

Apart from the standard uncertainty (u_s) that indicates the quality of direct measurements (e.g., the PTO displacement measured by a laser sensor), it is important to estimate the uncertainty relative to combined measurements, for example assessing the global uncertainty relative to the case of assessing the absorbed mechanical power by the PTO system, which is measured by multiple sensors. For this scope, the combined uncertainty u_b can be used.

The u_b value can be estimated firstly by defining a data reduction equation (DRE), which is an equation describing the indirect quantity evaluated. Then u_b can be found by applying a first-order Taylor series approximation. Depending on the specific PTO system used, a correspondent formulation of $u_c(P)$ can be obtained by implementing the theory described in [9,31], where examples for calculating $u_c(P)$ relative to an OWC and a tidal turbine are provided, respectively.

The combined uncertainty (u_b) can be converted into the expanded uncertainty if a specific confidence level needs to be assumed [31]. The expanded uncertainty, U, can be calculated as:

$$U = k \cdot u_c \tag{10}$$

where u_c is the combined uncertainty and k the coverage factor corresponding to a confidence level/interval percentage value that can be identified from the Student's distribution table on a specific number of tests repetitions carried out. The Student distribution is useful to be considered when few repetitions are available; otherwise, Gauss normal distribution shall be used for defining the confidence interval. In general, a minimum of 10 repetitions of specific testes is enough. This number of repetitions allows adopting a confidence interval equal to 95%, which corresponds to a relatively low k value ($k = 2.23$), as for standard Student's t-distribution [32].

Overall, the following recommendations concerning the uncertainty analysis can be provided:

- For reducing Type A uncertainty, as many repetitions as feasible of calibration and actual tests should be done;
- For reducing Type B uncertainty, experimental set-up and equipment need to be improved or upgraded before carrying-out calibration;
- The evaluation of Type B uncertainty can be done by gathering detailed specifications of the equipment used and/or regression analysis;
- To assess uncertainties related to the PTO, typically, the combined uncertainty (u_c), should be obtained. Following recognized formal practices, a specific formulation needs to be derived;
- For allowing the smallest expanded uncertainty value (U), a certain minimum number of tests are required. It is recommended to choose a coverage factor f in advance so to better plan the number of repetitions required during calibration and actual tests.

4. Case Studies

Depending on the WEC technology, the PTO physical modelling work at laboratory scale may involve the implementation of diverse experimental set-ups, calibration and error estimation procedures. For this reason, this section aims to provide a brief overview of three different case studies, which positively may be of aid for future work on PTO physical modelling at the laboratory.

4.1. Case Study 1: Closed Control Loop PTO

An idealized point absorber WEC was studied with numerical models and through experimental campaigns at the Department of Naval, Ocean and Marine Engineering of the University of Strathclyde [26]. During experiments, a closed control loop type of PTO simulator was used. The theoretical point absorber device, illustrated in Figure 4, is composed by a spherical floater, a tether mooring line (pre-tensioned) and a PTO system, which can be represented by a spring and damper components.

Models of the point absorber were built and tested at 1:86 and 1:33 scales (Figure 5). For these models, the PTO system was represented by a stainless-steel spring and an electric servomotor with a tachometer, which functioned as a damper mechanism. As illustrated in Figure 6 (same for both models) the PTO simulator was installed outside of the water. The mooring line from the spherical buoy was passing through a bottom-mounted pulley, the motor worm (Figure 7) and attached to the spring. The PTO set-up included a load cell, for measuring the load at an upper point, and a laser sensor, measuring the axial displacement of the mooring line.

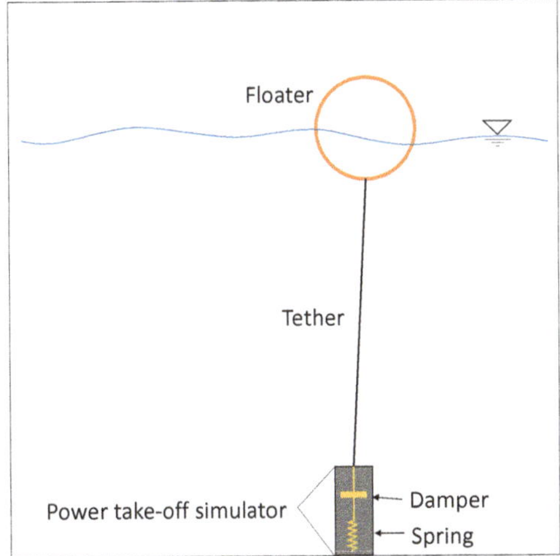

Figure 4. Spherical point absorber considered.

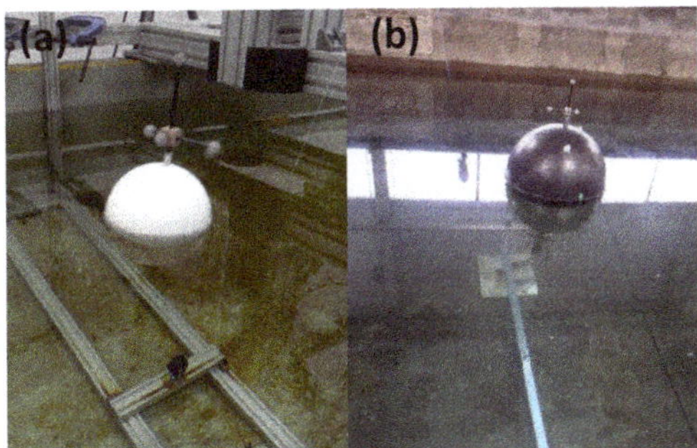

Figure 5. Point absorber devices tested at: (**a**) 1:86 and (**b**) 1:33 model scales.

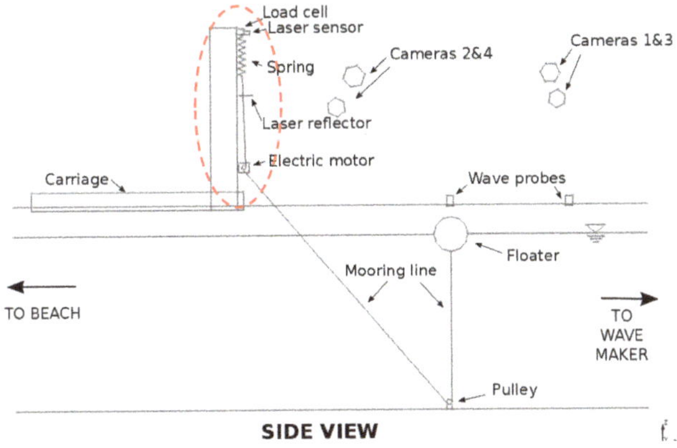

Figure 6. Experimental set-up for the point absorber experiments. The dashed red oval indicates the PTO simulator.

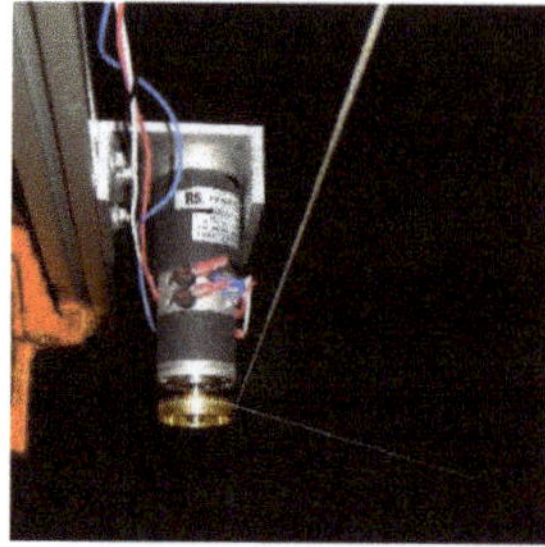

Figure 7. Electric servomotor and its worm damping the mooring line.

The calibration of the PTO was done in two stages. The first step consisted of dropping weights to drive the servomotor. By taking consideration of the falling speeds, measured over a vertical offset of 1 m and mass values of the weights, the damping forces exerted by the motor were calculated for a set of input current values. Following this stage, an initial calibration of the servomotor was obtained, Figure 8.

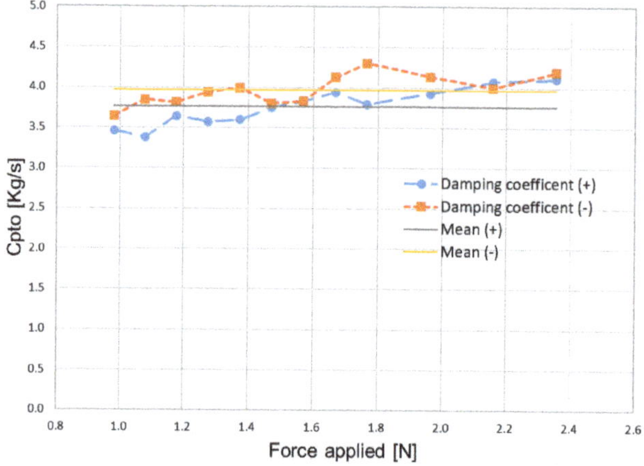

Figure 8. Initial calibration of PTO damping with weights drop-tests for the two directions of rotation of the servomotor.

However, the first stage resulted in not being accurate enough. This result was probably due to static and dynamic frictions and due to the inertia effects of the servomotor stator. Therefore, a second step of the calibration methodology was required for accurate tuning and characterization of the PTO simulator. This further stage was carried out by considering an oscillatory motion. Such motion is closer to the realistic behaviour of the PTO when the model is under the action of wave loads, respect to the linear motion relative to unidirectional weights-drop tests, which instead does not take into consideration accelerations and decelerations of the system. Thus, at the second stage, a set of calibration (dry) tests using a different rig (Figure 9) were performed. This rig is almost identical to the one later used as the final PTO simulator during wave tank tests.

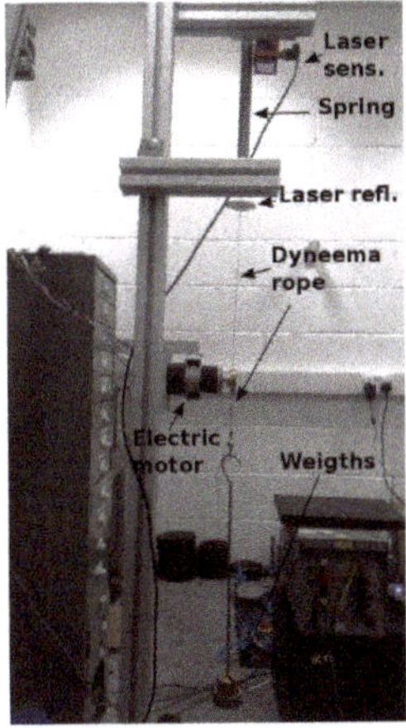

Figure 9. Calibration set-up for PTO simulator, adapted from [26].

Before defining the actual experimental set-up, the PTO simulator was, for an extended time, calibrated and characterized. For such a task, dry calibration tests were performed by assembling and using a preliminary calibration rig, Figure 9.

Calibration with oscillatory motion (Figure 9) can be carried out by using a suitable set of springs and weights needed for obtaining a relevant range of decay periods, which were initialized by an offset at $t = 0$. The classic equation of motion of the spring-damper harmonic oscillator was used for analysing the system:

$$m\ddot{z}(t) + C\dot{z}(t) + Kz(t) = 0 \tag{11}$$

where m is the mass, C the PTO damping coefficient, K the spring stiffness coefficient and z is the vertical displacement. By using a laser sensor, linked to a data-logging computer, it was possible to digitalize precisely the motion. Figure 10 shows an example of oscillations during an oscillatory calibration test.

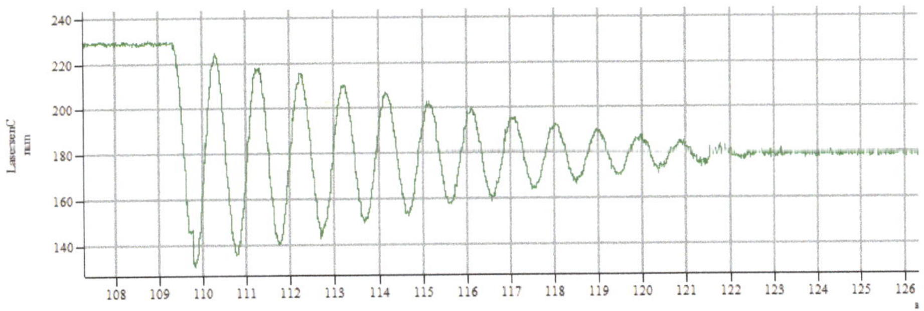

Figure 10. Example of decay test for servomotor calibration.

With the decay results and the aid of a simplified time-domain model based on Equation (11), the PTO damping values can be finely calibrated and the PTO system could be characterized by a matrix, which can have the format reported in Figure 11. This matrix could be used as a look-up table during experimental tests for better reproducing the PTO damping force.

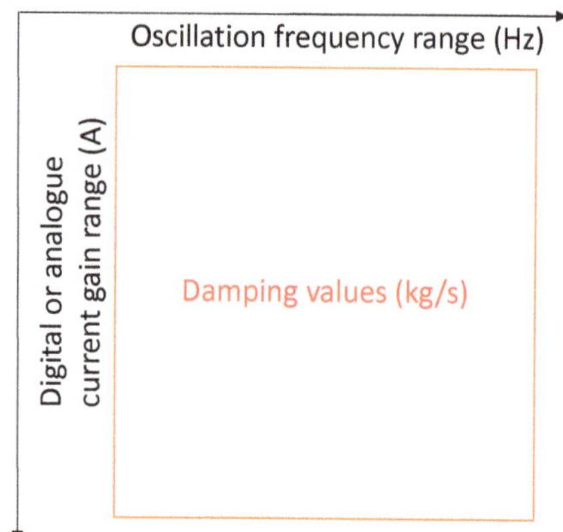

Figure 11. Calibration matrix of PTO simulator.

The final experimental set-up identified for the 1:33 scale model resulted to be appropriate for representing the dynamics and PTO behaviour of the point absorber. Besides, the two-stage calibration methodology developed allowed consistently reducing the uncertainties values [26].

4.2. Case Study 2: Eddy Current Based Electromagnetic PTOs

Two PTO models at a geometric scale of 1:30 have been designed at Uppsala University for a wave tank experiment. The main goals of the study were to develop a generic PTO testing solution, at a laboratory scale, and to study risks of failures and reliability of oscillatory types of WECs. Rotational and linear PTO models (Figures 12 and 13) were defined and assembled [33]. The PTOs are meant to be used within an experimental set-up consisting of a floating surface buoy connected by a line and pulley system to the PTO model, which is situated on a gantry above the water surface (similar model as for the previous case treated in Section 4.1).

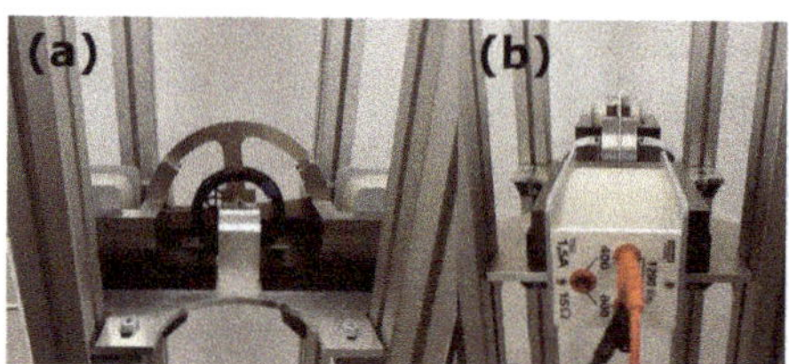

Figure 12. Rotational PTO, including the disk, the pulleys and the coils, (**a**) front and (**b**) lateral views.

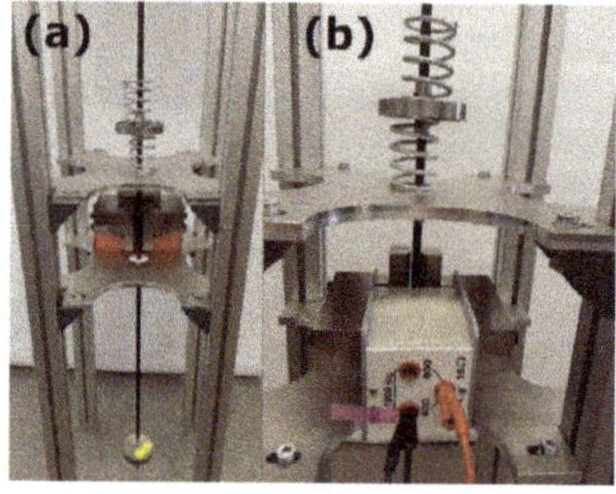

Figure 13. Linear PTO including the rod, the end-stop and the coil, (**a**) front and (**b**) lateral views.

The rotary PTO consists of a rotating aluminium disk mounted between two electromagnets that apply eddy current damping to the PTO. The magnetic field of the electromagnets is controlled by applying coil current in the range 0.2–1.4 A and the number of turns equal to 1200 for each coil. The motion of the buoy and line induces the rotational motion of the disk. During calibration procedures, clockwise and counter-clockwise rotations drop-weights tests have been conducted using weights in the range 0.2–1.2 kg. With the purpose of simplification, the end stop effect is not included in the rotary PTO. In this set-up, a plastic block stand can be used to minimize the flux dissipation to the structure and for concentrating the flux in the air gap, as shown in Figure 12. The rotary PTO employs eddy current braking for its damping force. The rotation of the disk in the electromagnetic field by the coils forms the eddy current on the disk resulting in a repulsive force (Lorenz force) between the disk and the coil [34]. The Lorenz force corresponds to the induced eddy current and magnitude of the magnetic field. The induced eddy current on the disk is proportional to the disk tangential velocity and magnitude of the magnetic field. Hence, the eddy current damping brake force relates to the square of the coil current. Increasing the current to enforce the magnetic field is applicable and effective only before the saturation of the coil core.

Similarly, in the linear PTO (Figure 13), the steel rod moves vertically in a magnetic field that is created by varying the coil current in the same range as rotary PTO but with 800 turns coil. As for the rotary PTO, variable weights are used; these are attached to the end of the rod for retraction as seen in Figure 13. For the linear PTO, the end stops effect is taken into consideration by two springs attached to the rod that confines the stroke displacement. A temporal magnet is formed in the rod due to the ferromagnetic properties of steel, i.e., resulting in non-permanent tension of the rod to one pole of the magnet (coil), which prompts pure damping force between the contact surface of the rod and one pole of the coil. A range of air gaps values was initially assessed for selecting the most suitable distance between the translating rod and the electromagnets (Figure 13b). This value allowed a wide range of desired damping forces. The linear PTO exerts a damping force, to the translatory motion of the buoy/line, which is proportional to the coil magnetic field.

For both PTOs, wire-draw-line position sensor with a spring constant of 0.6 N/m and linearity of ±0.25% of full-scale output (FSO) measured the vertical displacement of the rope attached to the PTOs and the weight. The data acquisition unit sampled the position data at 128 Hz.

To assess the performance of the two PTOs, dry tests with a range of attached weights can be performed. Depending on the mass of the attached weight, the dynamics of the system, the period of the drop and the number of data points varies. The heavier attached weight leads to higher inertia and a greater speed of the drop, resulting in a larger variance in damping estimation due to the few sampled data points available. On the other hand, the lighter attached weight results in friction domination due to the bearings, lower velocity. In this case a higher number of data points and, consequently, a lower variance in the estimation occur. Therefore, it emphasizes the importance of repetition of dry testing with a multitude of various weights to reach an appropriate level of bias and variance in estimating the damping value.

Figure 14 displays results of the damping for the rotary PTO, where the damping coefficient rises by increasing the current and it reveals minor fluctuation with the mass of the weights. The maximum damping force attains 2.9 Ns/m in the scale model, which corresponds to a real scale value of 14.3 kNs/m. The converging increase of the pattern of the damping coefficient by the current to a constant value indicates the coil's core saturation.

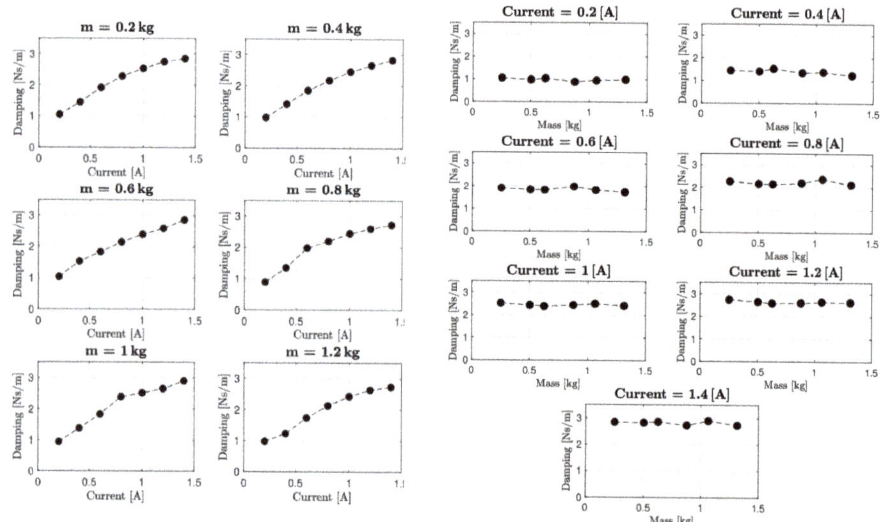

Figure 14. Results of rotary PTO for scale 1:30. To the right is the damping versus the mass of the weight for different coil current. To the left is the damping versus current for a different mass of the weight.

Figure 15 depicts the result of the dry testing for the linear PTO, where the highest friction force of 10.5 N is obtained for the scaled PTO model, corresponding to 283.5 kN in the full scale.

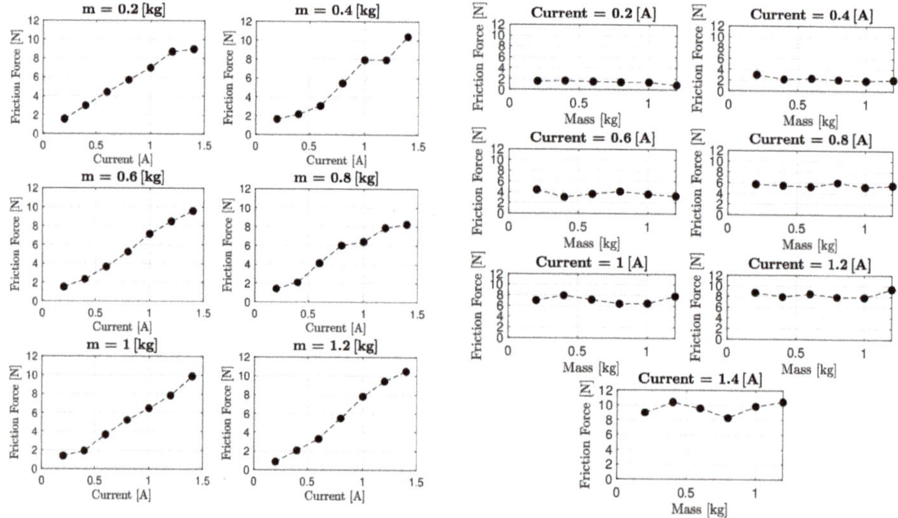

Figure 15. Results of linear PTO for 1:30 scale. To the right is the friction force versus the mass of the weight for different coil current. To the left is the damping versus current for a different mass of the weight.

To achieve optimal damping for power maximizing purposes, a higher damping value is required. This can be accomplished by increasing the number of pairs of magnets. Increasing the number of winding or coil current would have minimal influence on the eddy damping since the coil cores seem to reach its saturation through the maximum magnetic field strength (H) obtained here. Saturated coil confines the enhancement of the damping coefficient by limiting the magnetic flux in the core.

The non-contact eddy current damper depicts satisfactory design characterization. Nonetheless, to achieve higher damping some change of practice as mentioned is necessary. The linear PTO damper represents a simple and robust system that can be useful as a PTO simulator in the wave tank environment. For more information about the experimental set-up, the reader can refer to Shahroozi et al. [33].

4.3. Case Study 3: CECO Experimental PTO Physical Model

The CECO device is a sloped type of oscillatory WEC having a direct drive type of PTO system. This device is composed of two lateral mobile modules (LMMs), a central frame, and a fixed supporting structure, Figure 16a. The block formed by the central frame connected to the two LMMs oscillates along an inclined direction of motion, Figure 16b. Different scaled models (1:20–25) of CECO were constructed and tested at the wave basin of the Hydraulics Laboratory of the Faculty of Engineering of the University of Porto [35–37]. As envisioned for the real-scale design, the frame motions activate the PTO system enclosed into a fixed or floating supporting structure [38,39].

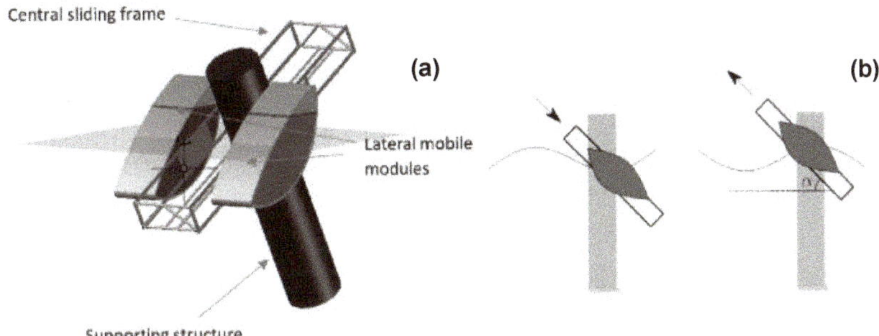

Figure 16. CECO wave energy converter: (**a**) main components and (**b**) working concept.

The PTO of CECO mainly consists of a rack-pinion mechanism and an electric generator, Figure 17b. The frame is constrained to slide along the inclined direction by a set of low friction bearings, Figure 17c.

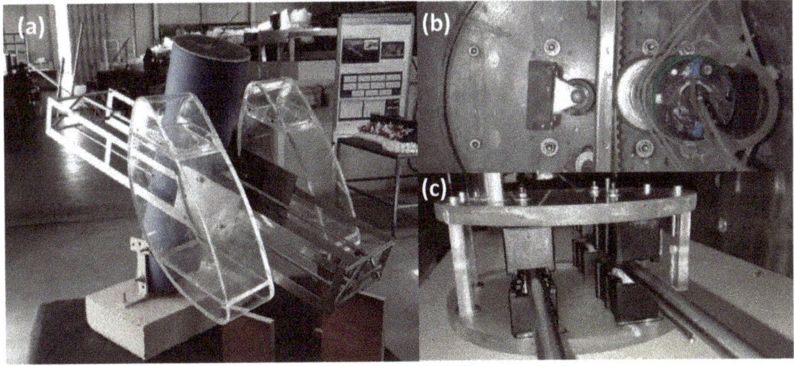

Figure 17. CECO wave energy converter: (**a**) 1:25 physical model, (**b**) rack-pinion system, (**c**) sliding frame fittings.

The experimental models of CECO were used in proof-of-concept testing. The PTO generator was able to simulate realistically the damping effect. For that purpose, an electric circuit was built and used for implementing different external resistances (in the range of 1 to 100 Ω) as generator load. The method for characterizing the PTO was based on the following equation [36]:

$$\overline{P}_{pto} = P_{T\_0} - P_{T\_1} \tag{12}$$

where P_{T_0} and P_{T_1} represent the power absorbed without and with the PTO damping, respectively. The wave power absorption can then be assessed by considering the model-scale device as a kinetic energy harvester (KEH). Based on the results of regular wave tests, a series of PTO damping values can be defined (C_{pto}) associated with the resistance values applied to the PTO circuit. In combination, the response amplitude operators (RAO) and the energy spectrum of response can be calculated. Successively, assuming linearity, the power in irregular sea states can be estimated by integrating the following equation [36]:

$$\overline{dP}_{pto}(f) = m\omega^3 \sqrt{S_\zeta(f)} \left(\sqrt{S_0(f)} - \sqrt{S_1(f)} \right) df \tag{13}$$

where m is the mass of the LMMs, ω the angular frequency of the LMMs oscillations and S_ζ the sea spectrum. S_0 and S_1 are the energy spectrum of response for the device without and with the PTO simulator switched on, respectively.

The method applied for characterizing the PTO of CECO based on the KEH approach allows easily overcoming uncertainties due to the unwanted energy losses related to the dynamic or static friction existing between the moving components of the experimental set-up. The technique is valuable at an initial stage of research, for instance, during the preliminary studies, where pilot assessments of the device efficiency and the concept functionality have to be carried out.

4.4. Discussion

The case studies treated allowed to confirm that if certain recommendations and approaches are being followed the experimental testing can be improved and carried out successfully. Each case study had a different way of PTO modelling thus eventually some of the recommendations depending on a particular case may be less relevant to others. In Table 3 are summarized recommendations and approaches proposed. In this table, the symbol "V" indicates that the mentioned recommendation is documented within the correspondent case study and the "Na" symbol, instead, means that the recommendation is not explicitly adopted within the study or no information is available.

Table 3. Recommendations implemented during case studies.

	Recommendation or Approach	Case Study 1	Case Study 2	Case Study 3
Experimental set-up	Minimize friction	V	V	V
	Minimize inertia of PTO components	Na	Na	V
	Use of industrial-grade equipment	V	V	Na
	Ensure rigidity	V	Na	Na
	Use machined parts and advanced materials	V	V	V
	Reduce complexities	V	V	Na
Calibration	Explore target PTO force and velocities values	V	V	V
	PTO oscillation tests	V	Na	Na
	Test, disassemble, reassemble, and re-test	V	Na	V
	Keep the PTO rig as it is when moving in the wave tank	V	Na	V
Errors estimat.	Informal error estimation	Na	Na	V
	Formal uncertainty analysis methods	V	Na	Na

The three case studies address and attempt to overcome specific challenges related to PTO physical modelling. The first case study aimed at reducing friction by trying to improve calibration methods and the experimental set-up, including all its components, e.g., measurements sensors, the electric motor, the spring, the pulley, and mooring cable. During the study, the uncertainty analysis allowed us to perform valid improvements of the experimental set-up and PTO calibration. Differently, the second case study focused on an eddy current based PTO damping mechanisms. Such custom-made PTO simulators allow very low friction to occur and have the advantage that can be highly tailored to needs compared to using an electric motor (e.g., as for Case study 1). As with this type of PTO simulator high accuracy may be expected, the option could be optimal. On the other hand, dealing with tolerances, the PTO system design and tuning can be laborious processes. Lately, the third case study points out that a miniaturized type of PTO can still be used at small scales with some degree of uncertainties. This case study also showed that if a KEH approach is implemented in assessing experimental results obtained with the adopted type of PTO physical model, a good insight into the device performance characteristics could be achieved.

5. Conclusions

In this paper, an overview of PTO physical modelling for WEC testing at a small scale was presented. Efforts were directed towards providing basic theory, practical guidelines, recommendations and examples for planning PTO physical modelling.

To obtain reliable experimental results, attention should be paid towards adopting and choosing the correct experimental set-up, calibration and uncertainty analysis practices. Concerning the experimental set-up, particular focus should be oriented for reducing mechanical friction between moving parts. Once the experimental set-up is optimized, extended calibration, work is required. Eventually, the implementation of more than a single methodology for characterizing and calibrating the PTO system could lead to better results. This observation is supported by examples relative to previous experimental works, at model scale, concerning three different WECs, which were treated in the form of case studies. In general, depending on the type of system, reproducing the effect of the PTO at the laboratory may be done by implementing only customized solutions. Nevertheless, methodologies applied in the past by other authors (e.g., those covered in the case studies presented) can be eventually be adopted in future experiments. In general, it is also of utmost relevance to plan well, with some anticipation, the experimental work and carefully document the experimental set-up, procedures, and calibration methodologies undertaken. Finally, to support results it is important to assess accurately PTO related uncertainty. In this context, a formal uncertainty analysis is highly suggested at the different stages of progress.

Author Contributions: Conceptualization, G.G., I.T. and P.R.-S.; methodology, G.G.; formal analysis, G.G.; investigation, G.G. and Z.S.; resources, G.G., Z.S. and P.R.-S.; data curation, G.G. and Z.S.; writing—original draft preparation, G.G. and Z.S.; writing—review and editing, P.R.-S., I.T., M.G., J.E. and V.R.; visualization, G.G. and Z.S.; supervision, P.R.-S., I.T., M.G., S.D. and J.E.; project administration, P.R.-S. and F.T.-P.; funding acquisition, G.G., P.R.-S. and F.T.-P. All authors have read and agreed to the published version of the manuscript.

Funding: This work was performed as part of a Short Term Scientific Mission (STSM) at Uppsala University, Sweden, undertaken by the first author, thank the WECANet COST Action (CA17105). The authors would like also to thank the support from: the Project OPWEC (POCI-01-0145-FEDER-016882, PTDC/MAR-TEC/6984/2014) funded/co-funded by FEDER through COMPETE 2020–Programa Operacional Competitividade e Internacionalização (POCI) and by Portuguese national funds, through the FCT-Fundação para a Ciência e a Tecnologia, IP; the project PORTOS–Ports Towards Energy Self-Sufficiency (EAPA 784/2018), co-financed by the Interreg Atlantic Area Programme through the European Regional Development Fund; and the project WEC4Ports–A hybrid Wave Energy Converter for Ports (OCEANERA-NET COFUND) funded under the frame of FCT. The authors would like to further acknowledge the project STandUP for Energy and Project 47264-1 funded by the Swedish Energy Authority, and the support provided by the Uppsala University.

Acknowledgments: The first author would like to acknowledge the Department of Naval, Architecture, Ocean and Marine Engineering of the University of Strathclyde for the support during his PhD.

Conflicts of Interest: The authors declare no conflict of interest.

References

1. De O. Falcão, A.F. Wave energy utilization: A review of the technologies. *Renew. Sustain. Energy Rev.* **2010**, *14*, 899–918. [CrossRef]
2. Cruz, J. *Ocean Wave Energy Current Status and Future Perspectives*; Springer Science & Business Media: Bristol, UK, 2008.
3. Sheng, W.; Alcorn, R.; Lewis, T. Physical modelling of wave energy converters. *Ocean Eng.* **2014**, *84*, 29–36. [CrossRef]
4. Beatty, S.; Ferri, F.; Bocking, B.; Kofoed, P.J.; Buckham, B. Power Take-Off Simulation for Scale Model Testing of Wave Energy Converters. *Energies* **2017**, *10*, 973. [CrossRef]
5. International Towing Tank Conference. Uncertainty Analysis, Instrument Calibration. Available online: https://www.ittc.info/media/7979/75-01-03-01.pdf (accessed on 24 March 2020).
6. International Towing Tank Conference. Testing and Extrapolation Methods Resistance Test. Available online: https://ittc.info/media/2019/75-02-02-01.pdf (accessed on 24 March 2020).
7. Marine Renewables Infrastructure Network. WP2: Marine Energy System Testing—Standardisation and Best Practice Deliverable 2.11 Best Practice Manual for PTO Testing. 2015. Available online: http://www.marinet2.eu/wp-content/uploads/2017/04/D2.11-best-practice-manual-for-PTO-testing.pdf (accessed on 24 March 2020).
8. Payne, G. Guidance for the Experimental Tank Testing of Wave Energy Converters. Available online: https://www.supergen-marine.org.uk/sites/supergen-marine.org.uk/files/publications/WEC_tank_testing.pdf (accessed on 24 March 2020).
9. International Towing Tank Conference. Uncertainty Analysis for a Wave Energy Converter. Available online: https://www.ittc.info/media/8135/75-02-07-0312.pdf (accessed on 24 March 2020).
10. European Marine Energy Cetre. Tank Testing on Wave Energy Conversion Systems. 2009. Available online: http://www.emec.org.uk/ (accessed on 2 March 2020).
11. Viviano, A.; Naty, S.; Foti, E. Scale effects in physical modelling of a generalized OWC. *Ocean Eng.* **2018**, *162*, 248–258. [CrossRef]
12. Falcão, A.F.O.; Henriques, J.C.C. Model-prototype similarity of oscillating-water-column wave energy converters. *Int. J. Mar. Energy* **2014**, *6*, 18–34. [CrossRef]
13. Benreguig, P.; Murphy, J. Modelling Air Compressibility in OWC Devices with Deformable Air Chambers. *J. Mar. Sci. Eng.* **2019**, *7*, 268. [CrossRef]
14. Falcão, A.F.O.; Henriques, J.C.C. Oscillating-water-column wave energy converters and air turbines: A review. *Renew. Energy* **2016**, *85*, 1391–1424. [CrossRef]
15. Fleming, A.; Macfarlane, G. In-situ orifice calibration for reversing oscillating flow and improved performance prediction for oscillating water column model test experiments. *Int. J. Mar. Energy* **2017**, *17*, 147–155. [CrossRef]
16. Xia, J.; Durfee, W.K. Analysis of Small-Scale Hydraulic Actuation Systems. *J. Mech. Des.* **2013**, *135*, 091001. [CrossRef]
17. Jin, S.; Patton, R.J.; Guo, B. Viscosity effect on a point absorber wave energy converter hydrodynamics validated by simulation and experiment. *Renew. Energy* **2018**, *129*, 500–512. [CrossRef]
18. Hinostroza, M.A.; Xu, H.T.; Guedes Soares, C. Manouvring test for a self-running ship model in various water depth conditions. In Proceedings of the 18th International Congress of the Maritme Association of the Mediterranean (IMAM 2019), Varna, Bulgaria, 9–11 September 2019.
19. Payne, G.S.; Taylor, J.R.M.; Bruce, T.; Parkin, P. Assessment of boundary-element method for modelling a free-floating sloped wave energy device. Part 2: Experimental validation. *Ocean Eng.* **2008**, *35*, 342–357. [CrossRef]
20. Beatty, S.J.; Hall, M.; Buckham, B.J.; Wild, P.; Bocking, B. Experimental and numerical comparisons of self-reacting point absorber wave energy converters in regular waves. *Ocean Eng.* **2015**, *104*, 370–386. [CrossRef]
21. Ruehl, K.; Forbush, D.D.; Yu, Y.-H.; Tom, N. Experimental and numerical comparisons of a dual-flap floating oscillating surge wave energy converter in regular waves. *Ocean Eng.* **2020**, *196*, 106575. [CrossRef]
22. Zurkinden, A.S.; Ferri, F.; Beatty, S.; Kofoed, J.P.; Kramer, M.M. Non-linear numerical modeling and experimental testing of a point absorber wave energy converter. *Ocean Eng.* **2014**, *78*, 11–21. [CrossRef]

23. Vijayakrishna Rapaka, E.; Natarajan, R.; Neelamani, S. Experimental investigation on the dynamic response of a moored wave energy device under regular sea waves. *Ocean Eng.* **2004**, *31*, 725–743. [CrossRef]

24. Chaplin, R.V. Seaweaver: A new surge-resonant wave energy converter. *Renew. Energy* **2013**, *57*, 662–670. [CrossRef]

25. Neill, S.P.; Hashemi, M.R. Chapter 1-Introduction. In *Fundamentals of Ocean Renewable Energy*; Neill, S.P., Hashemi, M.R., Eds.; Academic Press: London, UK, 2018; pp. 1–30. ISBN 978-0-12-810448-4. [CrossRef]

26. Giannini, G. *Mooring Analysis and Design for Wave Energy Converters*; University of Strathclyde Online Library: Glasgow, UK, 2018.

27. Taylor, J.R.M.; Mackay, I. The Design of an Eddy Current Dynamometer for a Free-Floating Sloped IPS Buoy. Available online: https://pdfs.semanticscholar.org/4cb4/8aa54cd7bfb8583245e3f3d05cefd36d84a1.pdf (accessed on 24 July 2020).

28. Lopes, M.F.P.; Henriques, J.C.C.; Lopes, M.C.; Gato, L.M.C.; Dente, A. Design of a non-linear power take-off simulator for model testing of rotating wave energy devices. In Proceedings of the 8th European Wave and Tidal Energy Conference (EWTEC), Uppsala, Sweden, 7–10 September 2009.

29. Troch, P.; Stratigaki, V.; Stallard, T.; Forehand, D.; Folley, M.; Kofoed, J.P.; Benoit, M.; Babarit, A.; Gallach-Sánchez, D.; De Bosscher, L.; et al. Physical Modelling of an Array of 25 Heaving Wave Energy Converters to Quantify Variation of Response and Wave Conditions. In Proceedings of the 10th European Wave and Tidal Energy Conference, Aalborg, Denmark, 2–5 September 2013.

30. Stratigaki, V.; Troch, P.; Stallard, T.; Forehand, D.; Kofoed, P.J.; Folley, M.; Benoit, M.; Babarit, A.; Kirkegaard, J. Wave Basin Experiments with Large Wave Energy Converter Arrays to Study Interactions between the Converters and Effects on Other Users in the Sea and the Coastal Area. *Energies* **2014**, *7*, 701–734. [CrossRef]

31. EquiMar. Equitable Testing and Evaluation of Marine Energy Extraction Devices in terms of Performance, Cost and Environmental Impact. Available online: https://tethys.pnnl.gov/publications/equitable-testing-evaluation-marine-energy-extraction-devices-terms-performance-cost (accessed on 24 July 2020).

32. Beyer, W. *Handbook of Tables for Probability and Statistics*, 2nd ed.; CRC Press: Boca Raton, FL, USA, 2017.

33. Shahroozi, Z.; Eriksson, M.; Göteman, M.; Engström, J. Design and evaluation of linear and rotational generator scale models for wave tank testing. In Proceedings of the 4th International Conference on Renewable Energies Offshore (RENEW), Lisbon, Portugal, 12–15 October 2020.

34. Sodano, H.A.; Bae, J.-S. Eddy Current Damping in Structures. *Shock Vib. Dig.* **2004**, *36*, 469. [CrossRef]

35. Rodríguez, C.A.; Rosa-Santos, P.; Taveira-Pinto, F. Hydrodynamic optimization of the geometry of a sloped-motion wave energy converter. *Ocean Eng.* **2020**, *199*, 1070468. [CrossRef]

36. Rodríguez, C.A.; Rosa-Santos, P.; Taveira-Pinto, F. Experimental Assessment of the Performance of CECO Wave Energy Converter in Irregular Waves. *J. Offshore Mech. Arct. Eng.* **2019**, *141*. [CrossRef]

37. Rosa-Santos, P.; Taveira-Pinto, F.; Teixeira, L.; Ribeiro, J. CECO wave energy converter: Experimental proof of concept. *J. Renew. Sustain. Energy* **2015**, *7*, 061704. [CrossRef]

38. Marinheiro, J.; Rosa-Santos, P.; Taveira-Pinto, F.; Ribeiro, J. Feasibility study of the CECO wave energy converter. In *Maritime Technology and Engineering*; Informa UK Limited: London, UK, 2014; pp. 1259–1267. ISBN 978-1-138-02727-5. [CrossRef]

39. Giannini, G.; Rosa-Santos, P.; Ramos, V.; Taveira-Pinto, F. On the Development of an Offshore Version of the CECO Wave Energy Converter. *Energies* **2020**, *13*, 1036. [CrossRef]

Journal of
Marine Science and Engineering

Article

Combined Floating Offshore Wind and Solar PV

Mario López [1], Noel Rodríguez [1] and Gregorio Iglesias [2,3,*]

[1] DyMAST Research Group & Department of Construction and Manufacturing Engineering, University of Oviedo, EPM, C/Gonzalo Gutiérrez Quirós s/n, 33600 Mieres, Asturias, Spain; mario.lopez@uniovi.es (M.L.); noelrodriguez5@gmail.com (N.R.)
[2] MaREI, Environmental Research Institute & School of Engineering, University College Cork, College Road, P43 C573 Cork, Ireland
[3] School of Engineering, University of Plymouth, Drake Circus, Plymouth PL4 8AA, UK
*   Correspondence: gregorio.iglesias@ucc.ie; Tel.: +35-(321)-490-2523

Received: 8 July 2020; Accepted: 28 July 2020; Published: 30 July 2020

Abstract: To mitigate the effects of wind variability on power output, hybrid systems that combine offshore wind with other renewables are a promising option. In this work we explore the potential of combining offshore wind and solar power through a case study in Asturias (Spain)—a region where floating solutions are the only option for marine renewables due to the lack of shallow water areas, which renders bottom-fixed wind turbines inviable. Offshore wind and solar power resources and production are assessed based on high-resolution data and the technical specifications of commercial wind turbines and solar photovoltaic (PV) panels. Relative to a typical offshore wind farm, a combined offshore wind–solar farm is found to increase the capacity and the energy production per unit surface area by factors of ten and seven, respectively. In this manner, the utilization of the marine space is optimized. Moreover, the power output is significantly smoother. To quantify this benefit, a novel Power Smoothing (PS) index is introduced in this work. The PS index achieved by combining floating offshore wind and solar PV is found to be of up to 63%. Beyond the interest of hybrid systems in the case study, the advantages of combining floating wind and solar PV are extensible to other regions where marine renewable energies are being considered.

Keywords: marine renewable energy; wind energy; solar energy; resource assessment; hybrid energy systems

---

1. Introduction

The scarcity of habitable land, growing energy consumption and environmental repercussions of fossil fuels are fostering the development of renewable energy projects in the marine environment. The oceans receive 70% of the global primary energy resource: radiation from the sun [1]. Intensive research is devoted to developing technologies in offshore wind, wave and tidal energy as the main forms of marine renewable energy [2].

Regarding wave energy, there exists a vast resource with a high energy density and good predictability, two properties of major interest for electricity generation [3]. For this reason, many wave energy conversion concepts have been proposed during the past decades. Most of the wave energy converters can be grouped into one of the following categories: oscillating systems (e.g., CECO [4]), overtopping devices (e.g., Wavecat [5]) and oscillating water columns (e.g., [6]). However, despite the large research effort and number of available concepts, no technology appears to be mature enough at this point for commercial projects [7].

Tidal energy is another well-known marine renewable energy, which can be harvested by means of either tidal barrages or tidal stream turbines. Tidal barrages are a well-proven technology but have two major downsides—the large capital expenditure required, and the environmental impact [8].

As for tidal stream turbines, the locations of interest, i.e., where a project can be economically viable, are limited [9].

As for wind energy, offshore farms have been in operation and connected to the grid since the 1990's, and have experienced a substantial growth in the last decade, especially in Europe [10]. These commercial farms consist of wind turbines fixed to the sea bottom in water depths below 50 m by means of foundations, such as monopiles, gravity structures, jackets, tripods, and triples ([11], Figure 1). Nonetheless, because of the limited amount of shallow waters for deploying fixed foundations, most of the future offshore wind farms will be installed in deeper water [12]. For this purpose, wind turbines with floating foundations are required, with some concepts already tested under real conditions in the recent years (e.g., WindFloat [13]).

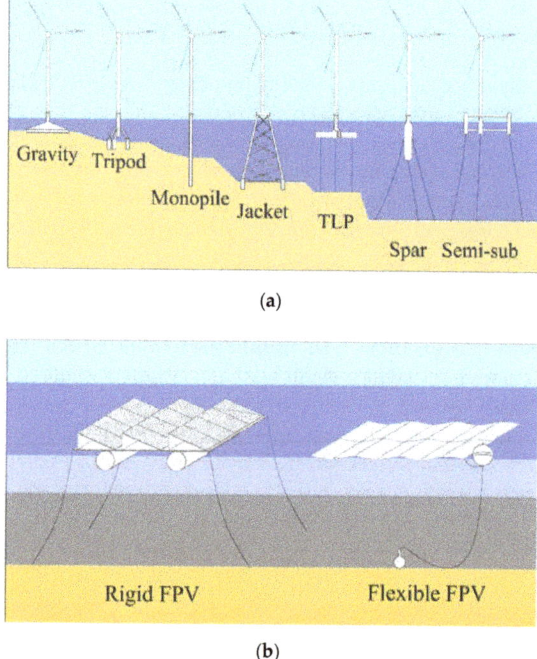

Figure 1. Different types of offshore wind turbine foundations (**a**) and floating photovoltaic (FPV) systems (**b**).

In addition to these marine renewable energy resources, there is an alternative that has been little explored in the marine environment: solar energy [14]. To harvest this resource in the oceans and seas, floating photovoltaic (FPV) systems are required (Figure 1). Although applying this technology in the marine environment is new, FPV farms have been deployed worldwide in freshwater, including lakes and reservoirs [15]. The main advantage of FPV systems against land-based ones is the water cooling on the solar cells [16]. This effect results in a higher energy conversion efficiency of the floating panels, which can generate up to 10% more electricity [17]. Other advantages of offshore FPV systems include: the availability of abundant water for cleaning the panels, the scalability of the systems from microwatt to megawatt, and the reduction in the growth of algae by the shading effect of the panels [14].

FPV systems in the marine environment can be more economical than wind farms at latitudes between 45° South and 45° North [18]. On these grounds, China and the Netherlands have started to deploy FPV systems in their maritime areas [19]. Notwithstanding, much work remains to be done in assessing the offshore photovoltaic potential (only a handful of areas investigated so far, e.g., India [20] and the Maltese islands [21]), and in developing reliable structures to resist the accelerated rusting in saltwater and the extreme dynamics of the marine actions (mainly winds, waves and tides) [22].

On another note, the combination of marine renewable energies is a promising solution that is supported by many synergies, such as the increase in the energy production and the reduction in the operation and maintenance cost [23]. Previous studies have shed light on the synergies between wind and wave energy, including topics such as the optimal array design (e.g., [24]) and the reduction in operation and maintenance costs (e.g., [25]). Nonetheless, wind–solar farm synergies remain unexplored, and only their combined use with aquaculture has been proposed [26]. Bearing in mind this scenario, the potential of combined wind–solar farms should be evaluated. A basic arrangement would be filling with FPV panels the free-surface amidst the offshore wind turbines, which avoids interferences in the production of both renewables (Figure 2).

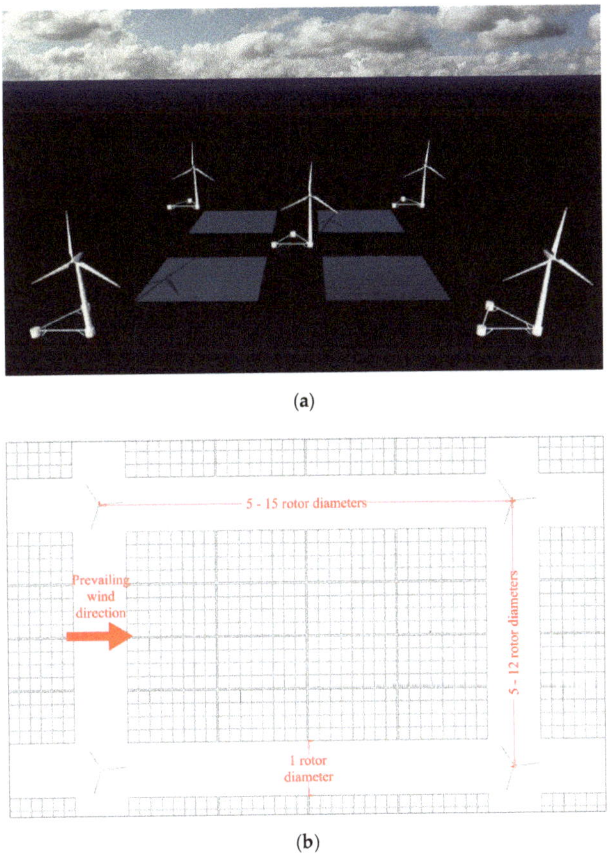

(a)

(b)

Figure 2. Combined floating wind and solar energy farm: general view (**a**) and schematic layout (**b**).

Asturias, a coastal region in Northern Spain with more than 300 km of coastline, is keen to develop its marine renewable energy potential. The wave energy resource in Asturias has been assessed [27]. Recently, Abanades et al. [28] proposed using wave energy converters in this region with a dual-purpose: the production of carbon-free energy and the mitigation of coastal erosion—a severe issue in the context of climate change. Regarding tidal energy, the available locations are scarce along the Asturian coastline. Some examples are the Ria of Ribadeo [29] and the Port of Aviles within the Nalon River estuary [9] (Figure 3). As for the offshore wind and solar energy resource, a detailed assessment is lacking. Although the narrow continental shelf compounds the deployment of fixed wind turbines, floating solutions may be used (Figure 1).

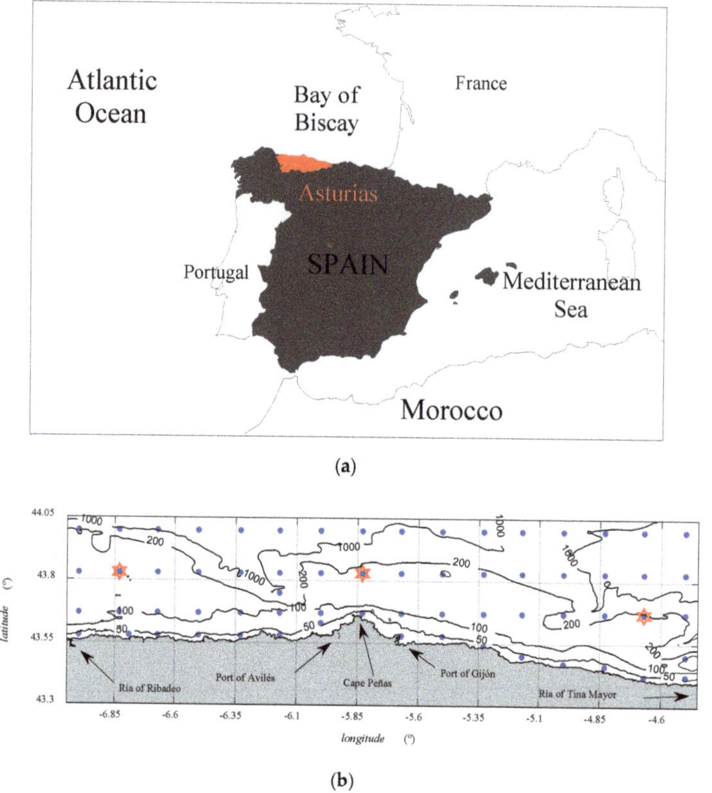

Figure 3. Location (**a**) and bathymetry (**b**) of Asturias. SIMAR data points are marked with blue dots, and the three study sites (West, Centre and East) with red stars. (Water depths in m).

In this work, the potential for developing floating wind and solar energy off the coast of Asturias is assessed and the benefits of their combination are examined. For this purpose, the gross resource of each renewable across the area of study is estimated by means of the available data, mainly from observational and numerical (hindcasting) meteorological databases. Then, the expected energy production of offshore solar and wind farms—separately and combined—is estimated considering the specifications of commercial technology. In the discussion, the potentials of offshore solar and wind energy are compared, taking into account the rated power of each technology, the marine space occupation, and the variability of their power output. From the results it is clear that combined offshore wind-solar farms present a production synergy that should be considered in future marine renewable energy projects.

The remainder of this paper is structured as follows: Section 2 describes the study area and presents the data used to assess the resources, the technology specifications used to calculate the energy production, and the methodology. In Section 3, results are presented and discussed. Finally, conclusions are drawn and further research lines suggested in Section 4.

2. Materials and Methods

2.1. Study Area

The area considered in this work is off the coast of Asturias, a region in North Spain between 43.3° and 44.1° North and 7.0° and 4.3° West (Figure 3). With only 10,603 km^2 of surface area, Asturias has 334 km of coastline, delimited by two estuaries: the Ria of Ribadeo to the west and the Ria of Tina Mayor to the east.

The climate of Asturias is of the oceanic type, with mild temperatures both in winter and summer, and the rainfall is abundant and well distributed throughout the year. The atmospheric circulation is governed by two centres of action, the Azores High and the Iceland Low, and more specifically by the North Atlantic Oscillation (NAO), i.e., the oscillation in the difference of atmospheric pressure between them. Because of this phenomenon, the winds are strongest in winter and lighter and less regular in summer. In summer, the poleward movement of the Azores High to about 35° North results in prevailing North East winds, which are cold and dry, bringing cool, clear, and rainless weather. The situation is different in winter, when the Azores High retreats to the south and allows a much more southerly trajectory of the Atlantic storms.

Wind energy areas can be grouped into classes from 1 to 7, with each class representing a range of mean wind power density or equivalent mean wind speed [30]. The study area falls in a transition zone between the energy-rich North Atlantic (with a wind power class up to 7) and the Bay of Biscay (with a wind power class below 4) [31]. On this basis, the study area may be considered a prospective development area for offshore wind energy. Bearing in mind the latitude of the study area (below 45°), FPV energy can be even more competitive than offshore wind energy according to [18]. It has been observed that more than 60% of the Asturian surface presents values lower than 3.4 kWh/(m^2·day) [32]. The sites with the highest values were reported to be in the coastal area around Cape Peñas, in accordance with studies of the neighbouring region of Galicia, which also found the highest irradiance to occur in coastal areas [33].

Regarding infrastructures and facilities for the installation, operation, maintenance and decommissioning of future offshore renewable energy farms, there are two major ports in Asturias: Gijón and Avilés, on both sides of Cape Peñas (Figure 3). Both present excellent maritime and terrestrial communications, which make them a hub for international trade, mainly with the Northern Europe (the ports are about 40 h from the North Sea) and the Americas. Moreover, both ports have infrastructures to manufacture, assemble and operate offshore renewable energy installations.

2.2. Spatial Data

To assess the offshore wind and solar energy resource in Asturias, two different sources of data were used. The datasets and parameters are presented in the subsequent sections.

2.2.1. SIMAR Dataset

SIMAR is an hourly dataset that covers the period from 1958 to the present by concatenating two subsets: SIMAR-44 and WANA. The SIMAR-44 subset covers the period 1958–1999 and is based on the joint numerical modelling of atmosphere, sea level and waves. Wind data are obtained by means of the regional model RCA3.5 (Rossby Center regional Atmospheric model) [34], which is fed with data from the re-analysis of meteorological observations produced by the European Centre for Medium-Range Weather Forecasts (ECMWF) in collaboration with many other institutions [35]. As for the WANA subset, it spans the period from 2000 to the present and obtains the wind fields with the High Resolution Limited Area Model (HIRLAM, [36]).

In this work, data from the 64 SIMAR data points shown in Figure 2 were considered. In particular, the time series of the mean wind speed (U_{10}) and mean wind direction (θ) at 10 m above the sea level were used. For a more detailed analysis, the data corresponding to three sites were used to characterize the West, Centre and East offshore regions of Asturias. These study sites correspond to the SIMAR data points nos. 3064044, 3088044, and 3116040, hereinafter referred to as points W, C and E, respectively, for the sake of simplicity (Figure 3).

2.2.2. POWER Dataset

To assess the offshore solar energy resource, solar radiation and air temperature data are required [37]. For this purpose, meteorological datasets from the Prediction of the Worldwide Energy Resources project (POWER) were collected [38]. POWER facilitates access to the satellite and modelling analyses of the National Aeronautics and Space Administration (NASA), which have been proved to be reliable and useful to the renewable energy sector and, especially, the solar energy industry [39]. The data in POWER are derived from the MERRA-2 assimilation model products [40] and the GEOS 5 near-real time products [41]. The MERRA-2 data spans the period from 1981 to within several months of real time. As for GEOS-5 dataset, it covers from the end of the MERRA-2 data stream to present.

Particularly, two daily time series were collected from POWER for this work: the all sky irradiance on a horizontal surface per day, R_{hor}, with units of W/m^2, and the average air temperature at 2 m, T_a, with units of °C. These time series were retrieved for the 64 data points in Figure 3.

2.3. Conversion Technology Overview

2.3.1. Offshore Wind Energy

In the marine environment, wind turbines require fixed or floating foundations depending on the operational water depth. Fixed foundations have been in use for over two decades, and include, in order of ascending water depth: gravity base, tripod, monopile and jacket-type (Figure 1). Floating foundations have started to be used more recently, in experimental windfarms such as Hywind and Windfloat, and target deeper waters, between 50 to 200 m [12]—the reason being their lower costs for construction, installation and decommissioning in this range of water depths [42]. There are three main groups of floating foundations, based on how the design achieves its stability [43]:

- tension leg platforms (or simply TLPs), in which a light structure is semi-submerged and anchored to the seabed through tensioned mooring lines for stability;
- spar buoys, in which a very large cylindrical buoy stabilizes the wind turbine using ballast (the center of gravity is much lower than the center of buoyancy), e.g., Hywind; and
- semi-submersible, in which the main principles of the two previous designs are combined, i.e., a semi-submerged structure is added to reach the necessary stability (e.g., Wind-Float).

In this work, three commercial three-bladed horizontal axis turbines with different values of rated power (P_R) and hub height were considered: the Senvion Repower 6.2 MW offshore (Siemens Gamesa, Zamudio, Spain), the Areva M5000 (Areva, Courbevoie, France) and the Siemens SWT-3.6-107 (Siemens Gamesa, Zamudio, Spain). Their power curves and technical specifications are shown in Figure 4 and Table 1, respectively [44]

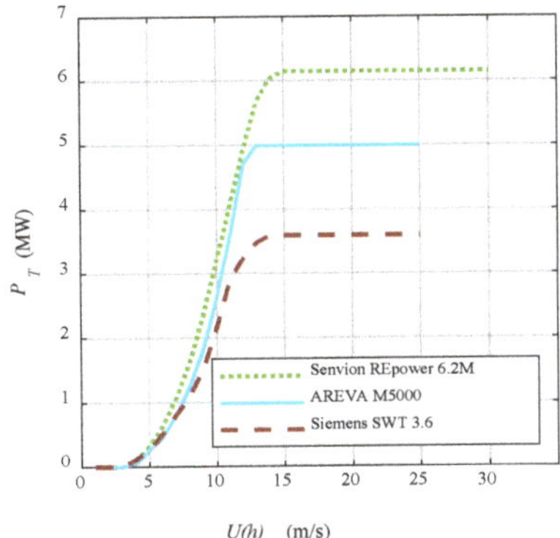

Figure 4. Power curves of the three wind turbines considered (data source: [44]).

Table 1. Technical specifications of the wind turbines considered (data source: [44]).

Parameter	Units	Wind Turbine		
		Repower 6.2M	M5000	SWT 3.6
P_R	MW	6.15	5.00	3.60
Rotor diameter	m	126.0	116.0	107.0
Hub height	m	95.0	90.0	88.0
Cut-in wind speed	m/s	3.5	4.0	4.0
Cut-off wind speed	m/s	30.0	25.0	25.0
Manufacturer	-	Senvion	Areva	Siemens

2.3.2. Offshore FPV Energy

FPV systems can be deployed in the marine environment by means of structures with some similarity to those installed to date in lakes and reservoirs [45]. Most of these systems consist of photovoltaic panels that are kept afloat by connecting them to (rigid) pontoons (Figure 2) [46]. These designs would require some modifications for the marine environment—considering the structural loadings induced by wave action—which would increase their cost significantly. For this reason, flexible systems floating on the waterline have been proposed (Figure 2) [47]. The aim of the flexible designs is to reduce the load on the structure and its mooring, without compromising its survivability in harsh marine environments [18].

As regards the light absorbing materials, there are five main groups of photovoltaic technologies: crystalline silicon, cadmium telluride and cadmium sulphide, organic and polymer cells, hybrid photovoltaic cells and thin film technology [48]. Crystalline silicon is usually considered for rigid pontoon systems (it is only available in a rigid format), whereas thin film systems are proposed for the flexible systems [18].

Only crystalline silicon panels were considered in this work, as it remains the dominant technology in photovoltaic power modules up to date [48]. More specifically, three commercial photovoltaic (PV) panels were considered (Table 2).

Table 2. Technical specifications of the solar photovoltaic (PV) panels considered (data source: [49–51]).

Parameter	Units	Panel Model		
		JKM 325PP-72V	**Tallmax TSM-DE**	**Q-Power-G5 280**
P_{STC}	W	325	365	280
Efficiency	%	16.75	18.8	17.1
α_p	$°\text{C}^{-1}$	−0.4	−0.39	−0.40
Length	m	1.96	1.96	1.65
Width	m	0.99	0.99	0.99
Surface	m^2	1.94	1.95	1.94
Weight	kg	26.5	26.0	22.2
Material	-	Si polycrystalline	Si monocrystalline	Si polycrystalline
Manufacturer	-	Jinko Solar	Trina Solar	Hanwa Q CELLS

2.4. Parameter Estimation and Definition

2.4.1. Wind Energy Assessment

It is well known that wind speed varies across the atmospheric boundary layer, and so does the energy resource. The wind speed profile across the boundary layer is mainly determined by the surface roughness characteristics, heat transfer and evaporation. Although complex wind profile expressions have been proposed, the most used wind speed profile—and the one used in this work—is the following [52]:

$$U(z) = U_{ref}\left(\frac{z}{z_{ref}}\right)^{\alpha},$$

(1)

where: U is the mean wind speed at height z above the sea surface; U_{ref} is the mean wind speed at the reference height z_{ref}; and α is an empirical coefficient that accounts for the site- and time-specific atmospheric conditions. Based on previous work for a nearby coastal region in the Cantabrian Sea, a value of $\alpha = 0.049$ was considered in this work [52]. Equation (1) was used to obtain the unknown value of U_T, the mean wind speed at the rotor height of a given wind turbine (z), from the known values of U_{10}, the mean wind speed at a reference height of $z_{ref} = 10$ m.

Prior to estimating the power output of the wind turbines, the time series of mean wind speeds were fitted to a Weibull distribution with the following probability density function [53]:

$$f(U|a,b) = \begin{cases} 0, & x < 0 \\ \frac{b}{a}\left(\frac{U}{a}\right)^{b-1}e^{-(U/a)^b}, & x \geq 0 \end{cases},$$

(2)

where U is the wind speed, and a and b are the shape and scale parameters, respectively. Accordingly, the cumulative density function of Weibull distribution is given by

$$F(U|a,b) = 1 - e^{-(U/a)^b}.$$

(3)

Having determined the Weibull distribution at a specific location and height above sea level, the average wind power density can be obtained as [12]

$$P_W = 0.5\rho_a \int U^3 f(U)dU,$$

(4)

where ρ_a is the air density.

The mean power output of a wind turbine at a specific location is obtained by combining the power curve of the turbine and the local wind speed distribution as follows [12]:

$$\overline{P}_{W,out} = P_T(U_T)f(U_T),$$

(5)

where $P_T(U_T)$ is the power curve defined as a density function of the mean wind speed at the hub height, and $f(U_T)$ is the Weibull probability density function of the mean wind speed at the rotor height. Accordingly, the energy output of a wind turbine in a given period of time ($E_{W,out}$) can be easily obtained by multiplying the mean power output by the duration of the period. Another factor that can be used to assess the performance of a wind turbine is the capacity factor [12], which is the ratio of the mean power output within the period of time considered to the rated power of the turbine (P_R),

$$CF_W = \frac{\overline{P}_{W,out}}{P_R}. \tag{6}$$

As a reference, in 2017 the capacity factors of all the offshore wind farms in Europe ranged from 29% to 48% [54]. Nonetheless, it should be noted that the capacity factor varies greatly depending on the location, the total available resource and the power curve of the offshore wind turbine considered, and thus values across multiple regions should be compared with caution [55].

2.4.2. Solar Energy Assessment

The temperature-corrected power output of a given FPV panel is given by [56]:

$$P_{S,out} = \eta P_{STC}\left(\frac{R_{hor}}{R_{STC}}\right)[1 - \alpha_P(T_m - T_{STC})], \tag{7}$$

where: η is a derating factor or performance ratio of the installation, which takes into account soiling of the panels, wiring losses, shading and aging, among other effects that reduce the efficiency of the system (absent shading, a representative value of $\eta = 0.85$ was considered); P_{STC} is the nominal power of the PV panel (i.e., its power output under Standard Test Conditions, STC); $R_{STC} = 1000$ W/m^2 is the irradiance at STC; α_P is the temperature coefficient of power, which depends on the PV module; T_m is the operational cell temperature; and $T_{STC} = 25\ ^{\circ}$C is the reference cell temperature at STC. The value of T_m for specific weather conditions and considering the water cooling of FPV panels can be obtained with the expression [17]:

$$T_m = e_0 + e_1 T_a + e_2 R_{hor} - e_3 U_{10}, \tag{8}$$

where $e_0 = 2.0458\ ^{\circ}$C, $e_1 = 0.9458\ ^{\circ}$C^{-1}, $e_2 = 0.0215\ ^{\circ}$C·m^2·day·kWh^{-1} and $e_3 = 1.2376\ ^{\circ}$C·s·m^{-1} are empirical parameters obtained by adjusting observational data from a real installation. The energy output of an FPV panel ($E_{S,out}$) can be simply obtained by integrating $P_{s,out}$ over time. As for the capacity factor of the FPV panel (CF_S), it is defined as the ratio of its average power output over time to its nominal power,

$$CF_S = \frac{\overline{P}_{S,out}}{P_{STC}}, \tag{9}$$

which, depending of the latitude of the offshore FPV system, varies between $CF_S = 8.6\%$ at the Poles and $CF_S = 20.5\%$ at the Equator for crystalline PV panels [18].

2.4.3. Specific Yield

The marine surface area occupied by the project should be considered when comparing different marine renewable energy technologies. For this purpose, the specific yield is defined as the energy output per unit surface area in an average year [18]; for floating wind turbines, it is calculated as

$$Y_W = E_{W,out}\frac{CD_W}{P_R}, \tag{10}$$

and for floating PV systems as

$$Y_S = E_{S,out}\frac{CD_S}{P_{STC}}, \tag{11}$$

where CD_W and CD_S are the capacity densities of wind and solar energy, respectively. The capacity density of a wind farm is defined as the ratio of the wind farm's rated capacity to its area, which includes the area occupied by the technology itself (footprint) and the area between the devices required to limit the wake effects or mooring/foundation interfaces. The optimal spacings between the offshore wind turbines usually vary between 5 and 15 rotor diameters along the prevailing wind direction, and between 5 and 12 rotor diameters along the crosswind direction [57].

Given the lack of real projects in the study area and bearing in mind the data from other European areas, a wind farm capacity density of $CD_W = 5.36$ MW/km^2 was considered in this work [58]. As in the case of wind farms, the capacity density of FPV farms is defined based on their occupation area. According to Trapani et al. [18], typical values range between $CD_S = 57$ and 74 MW/km^2 when the panels are installed at a nearly horizontal angle. An average value of $CD_S = 65$ MW/km^2 was chosen in this work.

2.4.4. Power Output Variability and Power Smoothing (PS) Index

The power output variability has been identified as a major cost driver in renewable energy projects. Moreover, a large variability can impede the penetration of offshore wind farms into the electricity market [58]. With the aim of achieving a smoother power output, diversified marine renewable energy farms have been proposed as a solution (e.g., combined wind-wave farms [59]).

The variability in the daily power output of a wind farm ($P_{w,out}$) or an FPV farm ($P_{S,out}$) is given by the coefficient of variation (CV_W or CV_S, respectively). To quantify the smoothing in the power output of offshore wind turbines due to their combination with FPV systems, a novel Power Smoothing (PS) index is proposed in this work,

$$PS = \frac{CV_W - CV_{WS}}{CV_W}, \qquad (12)$$

with CV_{WS} the coefficient of variation of the daily power output of a combined offshore wind–solar farm.

3. Results and Discussion

3.1. Offshore Wind Energy

3.1.1. Gross Resource

Figure 5 depicts the variations in the wind power density across the study area. Given that each model of offshore wind turbine presents a different hub height, the values were computed with Equation (4) and the mean wind speeds at the reference height (U_{10}). Values above 400 kW/m^2 (wind power class seven) are reached northwest of the study area, corresponding to locations with water depths above 200 m. In most of the areas with water depths between 50 and 200 m—where floating wind farms can be deployed—the wind power density ranges from 100 to 350 kW/m^2, corresponding to classes from two to six, in which wind energy development could be carried out [30].

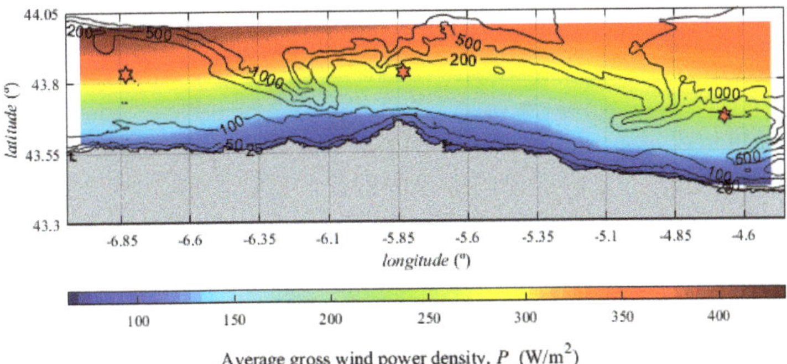

Average gross wind power density, P_w (W/m^2)

Figure 5. Average wind power density (P_w) off Asturias. The three study sites (West, Centre and East) are marked with red stars. (Water depths in m).

The results reveal three patterns in the spatial distribution of the resource (Figure 5). First, wind power density increases with distance from the coastline, from nearshore values below 100 W/m^2 to over 400 W/m^2 far offshore. Second, the resource decreases to the west and east of Cape Peñas, which may be explained by the sheltering effect of this headland. Third, the wind energy resource reduces slightly from West to East, which is apparent when comparing the wind speeds at the three study sites (Figure 6). The Weibull distributions peak at U_{10} = 4.30 and 3.35 m/s at the West and East sites, respectively.

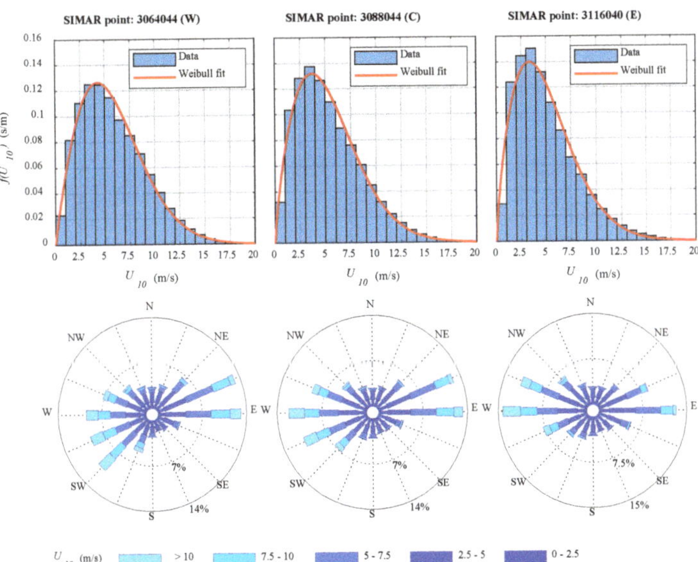

Figure 6. Probability Density Functions and wind roses at three study sites in the West, Centre, and East sections of the study area.

Regarding wind directions, both the dominant and prevailing wind directions present a westerly component—corresponding to the westerlies or anti-trades (Figure 6). As can be observed in the wind roses of the study sites, small differences occur due to the topographic effect of the continent nearby.

3.1.2. Performance Analysis

The energy output of the three wind turbines considered differed significantly across the area of study (Figure 7). The greatest production corresponded to the Senvion Repower 6.2 MW turbine, with values above 15 GWh/year for Northwestern locations. As expected, the spatial distribution of the energy output follows the pattern that was already described for the wind energy power density in Figure 5: the estimated energy output decreases towards the coast. Bearing in mind the high water depths off Asturias (Figure 3) and to maximize the energy output of the wind farms, floating foundations are required for the deployment of offshore wind turbines in the region.

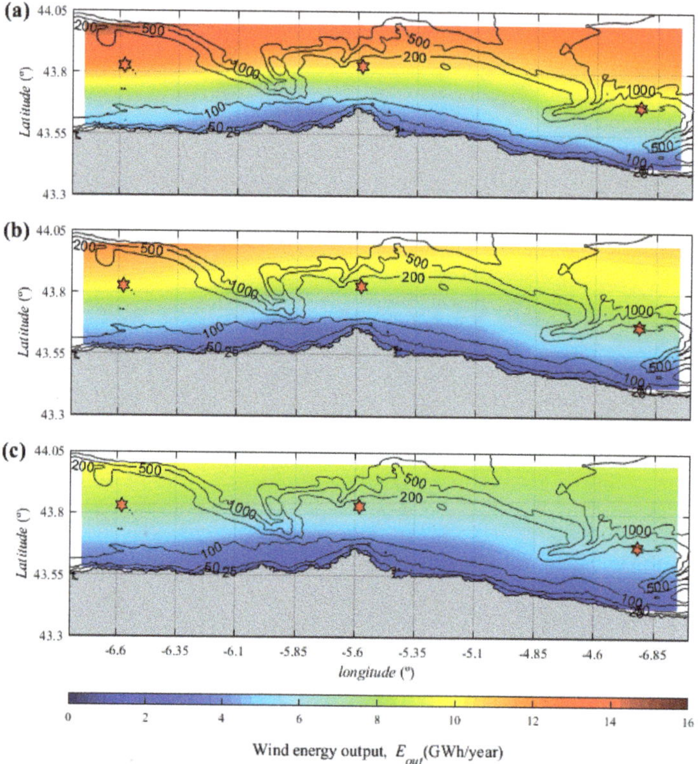

Figure 7. Expected wind energy output ($E_{w,out}$) for three commercial wind turbines: Senvion REpower 6.2M (**a**), Areva M5000 (**b**) and Siemens SWT 3.6 (**c**). The three study sites (West, Centre and East) are marked with red stars. (Water depths in m).

The capacity factor was calculated with Equation (6) for the three wind turbines at the three study sites (Table 2). It decreases with increasing rated power, and for a given turbine is maximum for the West study site. It is noticeable that, despite the high values obtained for the wind power density in the previous section, the capacity factor (between CF_W = 16% and 26%) turns out to be small when compared with the typical values for offshore wind energy in Europe (commonly above CF_W = 29%, [54]). The results in Table 3 are consistent with those reported for South West Portugal, where sites with similar levels of expected wind energy output ($E_{W,out}$ ~ 10 GWh/year) had capacity factors of CF_W = 25 and 26% for the Areva M5000 and the Senvion Repower 6.2M wind turbines, respectively [12].

Table 3. Capacity factor (*CF*, with units of %) of six commercial renewable energy technologies (three wind turbines and three PV panels) at the three study sites.

	Study Site		
	W	C	E
Senvion RE Power 6.2M	23.1	19.2	16.2
Areva M5000	23.0	19.7	16.8
Siemens SWT 3.6	25.9	22.4	19.4
Q-Power-G5 280	12.8	12.6	13.2
Tallmax TSM-DE	11.7	11.5	12.1
JKM 325PP-72V	12.2	12.0	12.5

3.2. Offshore FPV Energy

3.2.1. Gross Resource

The gross solar energy resource off the Asturian coast is mapped in Figure 8 based on the data obtained from the POWER datasets described in Section 2.2.2. POWER. The average solar irradiance varies across the study region, with values ranging from 125 to 165 W/m^2. The lowest values correspond to the vicinity of Cape Peñas and, especially, to eastern areas. The average irradiance increases from South to North within the study area (as expected for a region in the North Hemisphere) by about 10 W/m^2.

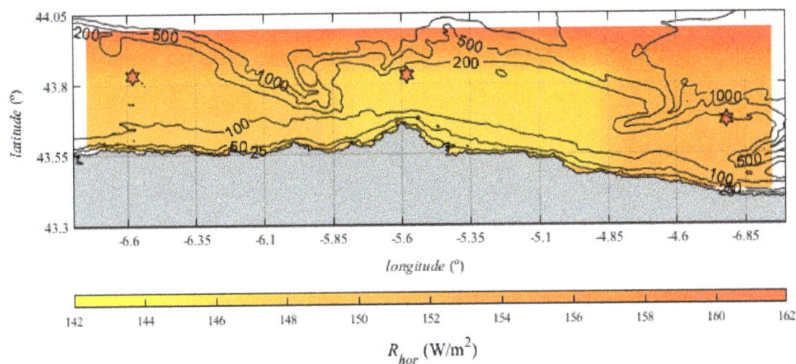

Figure 8. Average irradiance on a horizontal surface (R_{hor}) off Asturias. The three study sites (West, Centre and East) are marked with red stars. (Water depths in m).

3.2.2. Performance Analysis

The power output of the three PV panels considered in Table 2 was estimated by means of Equation (7) for each time step and data point in the area of interest. Then, the energy output in an average year was estimated considering the panels with a horizontal layout (i.e., with $\eta = 0.85$). The results are mapped in Figure 9.

Depending on the solar panel, the energy output varies between ~300 and ~400 kWh per average year across the region. The panel with the highest rated power and efficiency, the Tallmax TSM-DE (Trina Solar, Changzhou, China) produces the largest energy output in an average year (Table 1). As for the capacity factor, its values range between 12.2% and 13.2% with small differences between the three PV panels (Table 3).

The differences in the energy output are also small when comparing the results for a given PV panel at the three different study sites (W, C and E), showing that the influence of the water depth and/or the distance to the coast on the performance of the FPV systems across the study area is negligible. Therefore, there is no need in principle for deploying FPV farms in deep waters, apart from the availability of marine space or the synergies of joint deployment with, e.g., offshore wind turbines.

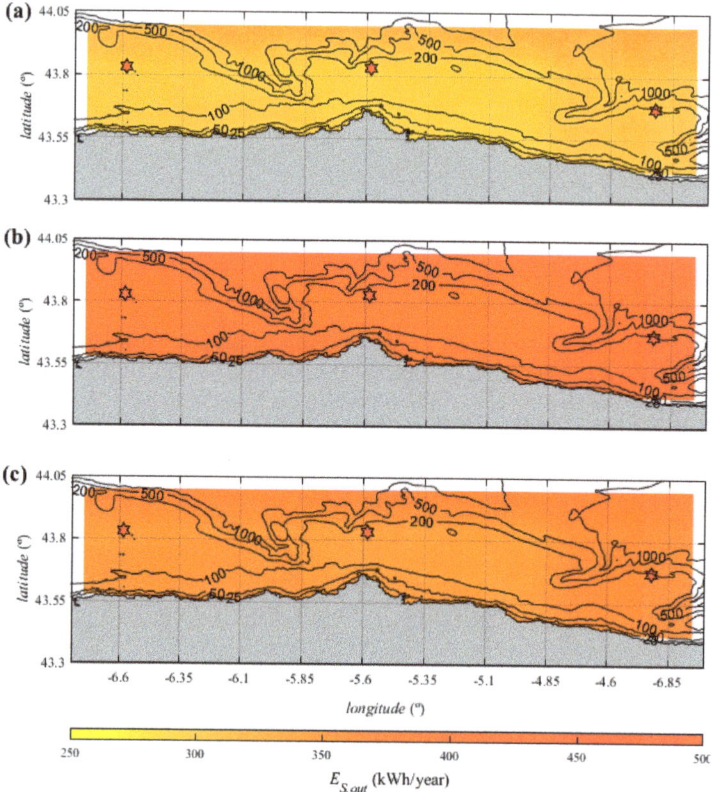

Figure 9. Expected solar energy output for three different commercial solar panels: Q-Power-G5 280 (**a**), Tallmax TSM-DE (**b**), and JKM 325PP-72V (**c**). The three study sites (West, Centre and East) are marked with red stars. (Water depths in m).

3.3. Comparative Analysis

Given the difference in rated power between a single wind turbine and a single PV panel (6.2 MW vs. 365 W, respectively), there is no point in comparing the energy production of single units. The apt comparison is between facilities with similar installed power. In this vein, the floating solar farm currently in operation at Fujian Zhangpu Zhuyu, China [19], with an installed power of 5 MW, was taken as a reference. The Senvion REPower 6.2M wind turbine and the Tallmax TSM-DE PV panel, which presented the best performance across the study area within their respective category, were selected as references for technical specifications (Figures 1 and 4).

The expected energy output of the solar farm at the three study sites is about half the expected energy output of the wind turbine (Table 4). As for the capacity factor, the values of the wind turbine are nearly twice as large as those of the FPV farm. This result would point in principle to a better performance of the wind turbines in comparison to FPV systems, but the variability of the power output and the area of the project should also be considered.

Table 4. Energy production (*E*), coefficient of variation (*CV*), specific yield (*Y*) and capacity factor (*CF*) for the offshore wind turbine and the offshore FPV farm.

			Study Site		
	Parameter	Units	W	C	E
6.2 MW offshore wind turbine	$E_{W,out}$	GWh/year	12.4	10.6	9.0
	CV_W	-	1.3	1.4	1.6
	$Y_{W,out}$	GWh/(km^2·year)	10.6	9.2	7.8
	CF_W	%	23.1	19.2	16.2
5 MW offshore FPV farm	$E_{S,out}$	GWh/year	5.3	5.2	5.5
	CV_S	-	0.6	0.6	0.6
	$Y_{S,out}$	GWh/(km^2·year)	69.4	67.6	71.5
	CF_S	%	11.7	11.5	12.1

The power output for the FPV farm presents a coefficient of variation $CV_S \approx 0.60$ for the three reference sites, which is half the wind turbine counterpart (Table 4). Based on the climate of the region, patterns can be recognized. The energy output of the wind turbine in winter doubles the summer values. The opposite holds for the FPV farm, with energy outputs in summer months up to three times higher. In fact, the energy output of the FPV farm increases to such an extent during summer that for the study site E it surpasses the energy production of the wind turbine (Figure 10).

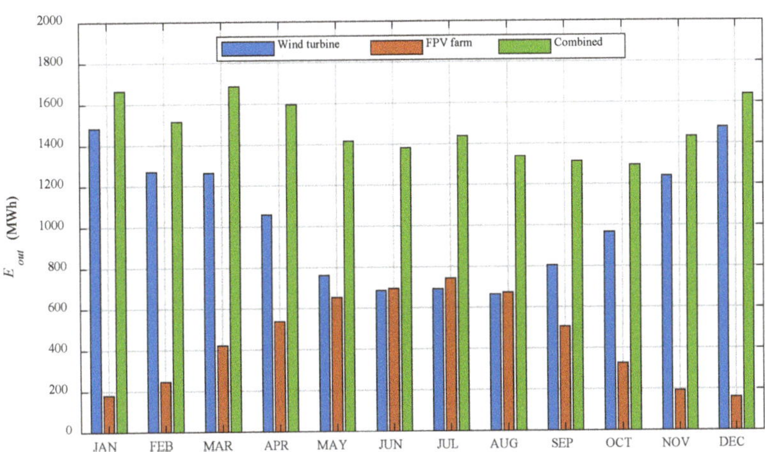

Figure 10. Monthly energy production at study site East of a 6.2 MW offshore wind turbine, a 5 MW FPV farm, and both combined.

The specific yield of each technology was obtained with Equations (10) and (11) for the three study sites. The results are summarized in Table 3. The values obtained for the offshore wind turbine across the study area vary between 7.8 and 10.6 GWh/km^2 per average year, whereas the values for the FPV farm vary little around ~70 GWh/km^2. Therefore, FPV systems would be nearly seven times more productive than offshore wind turbines for the same project area. This result is in part due to the fact that the area occupied by a wind farm includes "empty" spacing between turbines, whereas the area occupied by a solar project is all but entirely covered by the FPV systems.

3.4. Combined Offshore Wind and FPV Farm

Combining offshore wind turbines and FPV systems presents a series of advantages relative to conventional offshore wind farms. First and foremost, conventional offshore wind farms require large empty marine surface areas in between the turbines. In a combined farm, these surface areas are occupied with FPV systems which increases the capacity density. Consider a typical wind farm capacity density (CD_W = 5.36 MW/km^2) and the layout in Figure 2. The maximum surface area available for deploying FPV systems between wind turbines of the model Senvion Repower 6.2 MW (with a rotor diameter of 126 m) would vary between 64% and 86% of the total area, depending on the relationship between the turbine spacing in the prevailing wind direction and the turbine spacing in the crosswind direction (Figure 2). Accordingly, the share of FPV power in the combined offshore wind-solar farm would vary between 73% and 96% of the installed capacity. The capacity density of the combined farm would be ~57.5 MW/km^2 in all cases, which is ten times higher than the value of a conventional wind farm.

Regarding the capacity factor, a combined farm at study site E, for example, would present a value between 12.1% (pure offshore FPV farm) and 16.2% (pure offshore wind farm), depending on the share of FPV power in the total installed capacity (Figure 11). In terms of the specific yield, if the maximum typical spacing between the wind turbines is considered, a combined farm would produce a specific yield of up to 61.2 GWh/(km^2·year) at study site E—more than seven times the specific yield of wind turbines (Table 4). It follows that a combined offshore wind–solar farm can produce significantly more energy per surface unit area than an offshore wind farm.

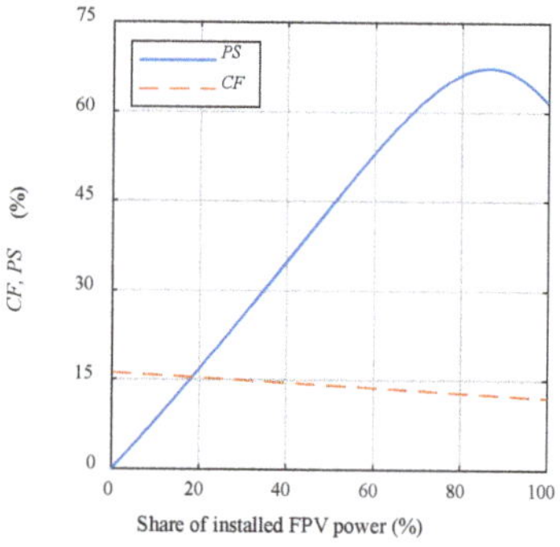

Figure 11. Capacity factor, *CF*, and Power Smoothing index, *PS*, for combined offshore wind and FPV farms as a function of share of installed FPV power capacity.

Another advantage is that the combination of FPV systems with offshore wind turbines significantly reduces the intra-annual variability of the energy output, which is one of the downsides of marine renewable energy [60]. This reduction is apparent in Figure 10. In the same line, hybrid systems present lower values of the coefficient of variability of power output. This is made apparent by the Power Smoothing (*PS*) index, which is plotted in Figure 11 as a function of the share of FPV installed power with respect to the total installed power of the hybrid farm at study site E. The PS index peaks at 68% for an FPV installed power share of 86%, implying a very substantial reduction in the variability of the power output with respect to the wind turbine. Moreover, the power output variability of the combined wind–solar farm (CV_{WS} = 0.51) is also significantly lower (~20%) than that of a stand-alone FPV farm (Table 4). It follows that hybrid systems combining FPV with offshore wind produce a smoother power output than conventional systems with either stand-alone wind or FPV—a significant advantage.

The latter results have been obtained without considering the issue of shading. Intermittent shadows from the rotors and towers of the wind turbines may affect the performance of the FPV systems.

4. Conclusions

The first research into combining offshore wind and solar power was conducted in this work through a case study off Asturias (North Spain). Floating technology would be required virtually anywhere in the study area, given that water depths exceed 50 m except in a narrow coastal fringe. The performance of several technologies—three wind turbines and three solar panels, separately and in combination—was examined by considering the energy output, the specific yield, the power output variability, and the capacity factor. Bearing in mind the results, a combined offshore wind–solar farm is proposed and its performance examined.

The wind energy resource across the study area decreases towards the coast, from ~400 W/m^2 in the deeper parts to ~100 W/m^2 near shore. The wind turbine with the highest rated power would yield 15 GWh/year in the North West section of the study area, which corresponds approximately to the limit of the continental shelf—where water depths increase abruptly from 200 m to over 1000 m.

As for the solar resource, the horizontal irradiance on a horizontal surface varies only weakly across the study area, with values of approx. 150 W/m^2, while the expected energy output for a commercial PV panel would vary between 300 to 400 kWh/year, depending on the model and the location. Unlike wind energy, the expected solar energy output varies little across the study area. Therefore, power plants in deep waters would not be required to maximize the solar energy output—an advantage of offshore PV farms versus offshore wind energy farms.

When the performance of a commercial 6.2 MW wind turbine and a 5 MW FPV farm are compared, the former presents far higher values of the energy output and the capacity factor. This first approximation could be considered favourable to wind energy, but the FPV farm presents a much lower variability of the production, and its specific yield is some seven times higher. These properties justify combining both with a view to smoothing the power output and realising economies of scale.

On this basis, a hybrid system was proposed and the production synergies were investigated. A combined offshore wind–solar farm can reach 57.5 MW/km^2 of capacity density and 61.2 GWh/(km^2·year) of specific yield—10 and 7 times the typical values for stand-alone offshore wind turbines, respectively. Furthermore, the power output of offshore wind turbines and FPV systems is significantly smoothed when both are combined, with a 68% reduction in the power output variability relative to a stand-alone wind farm.

In summary, a hybrid marine renewable energy farm with offshore wind turbines and FPV would not only increase the power output per unit surface area of marine space, but also improve the quality of the power output by reducing its temporal variability.

Author Contributions: Conceptualization, M.L., N.R. and G.I.; methodology, M.L., N.R. and G.I.; software, M.L. and N.R.; resources, N.R.; writing—original draft preparation, M.L.; writing—review and editing, G.I.; funding acquisition, M.L. and G.I. All authors have read and agreed to the published version of the manuscript.

Funding: This research was partially funded by the Ports Towards Energy Self-Sufficiency (PORTOS) project co-financed by the Interreg Atlantic Area Programme through the European Regional Development Fund, grant number EAPA_784/2018, and by the Council of Gijón through the University Institute of Industrial Technology of Asturias, grant number SV-20-GIJON-1-19 (the views and opinions expressed herein do not necessarily reflect those of the University Institute of Industrial Technology of Asturias—IUTA).

Acknowledgments: Wind datasets were provided by the Spanish government agency Puertos del Estado (Ministerio de Fomento). Irradiance datasets were obtained from the NASA Langley Research Center Atmospheric Science Data Center Surface meteorological and Solar Energy (SSE) web portal supported by the NASA LaRC POWER Project.

Conflicts of Interest: The authors declare no conflict of interest. The funders had no role in the design of the study; in the collection, analyses, or interpretation of data; in the writing of the manuscript, or in the decision to publish the results.

References

1. Pedersen, P.T. Marine Structures: Future Trends and the Role of Universities. *Engineering* **2015**, *1*, 131–138. [CrossRef]
2. Taveira-Pinto, F.F.; Iglesias, G.; Rosa-Santos, P.; Deng, Z.D. Preface to Special Topic: Marine Renewable Energy. *J. Renew. Sustain. Energy* **2015**, *7*, 5. [CrossRef]
3. Iglesias, G.; López, M.; Carballo, R.; Castro, A.; Fraguela, J.A.; Frigaard, P. Wave energy potential in Galicia (NW Spain). *Renew. Energy* **2009**, *34*, 2323–2333. [CrossRef]

4. López, M.; Taveira-Pinto, F.; Rosa-Santos, P. Numerical modelling of the CECO wave energy converter. *Renew. Energy* **2017**, *113*, 202–210. [CrossRef]

5. Fernandez, H.; Iglesias, G.; Carballo, R.; Castro, A.; Fraguela, J.A.; Taveira-Pinto, F.; Sanchez, M. The new wave energy converter WaveCat: Concept and laboratory tests. *Mar. Struct.* **2012**, *29*, 58–70. [CrossRef]

6. Veigas, M.; López, M.; Romillo, P.; Carballo, R.; Castro, A.; Iglesias, G. A proposed wave farm on the Galician coast. *Energy Convers. Manag.* **2015**, *99*, 102–111. [CrossRef]

7. Antonio, F.D.O. Wave energy utilization: A review of the technologies. *Renew. Sustain. Energy Rev.* **2010**, *14*, 899–918. [CrossRef]

8. Pacheco, A.; Ferreira, Ó.; Carballo, R.; Iglesias, G. Evaluation of the production of tidal stream energy in an inlet channel by coupling field data and numerical modelling. *Energy* **2014**, *71*, 104–117. [CrossRef]

9. Alvarez, E.A.; Rico-Secades, M.; Suárez, D.F.; Gutiérrez-Trashorras, A.J.; Fernández-Francos, J. Obtaining energy from tidal microturbines: A practical example in the Nalón River. *Appl. Energy* **2016**, *183*, 100–112. [CrossRef]

10. Esteban, M.D.; Diez, J.J.; López, J.S.; Negro, V. Why offshore wind energy? *Renew. Energy* **2011**, *36*. [CrossRef]

11. Esteban, M.D.; López-Gutiérrez, J.S.; Negro, V.; Matutano, C.; García-Flores, F.M.; Millán, M.Á. Offshore wind foundation design: Some key issues. *J. Environ. Eng.* **2015**, *141*, 1–6. [CrossRef]

12. Pacheco, A.; Gorbeña, E.; Sequeira, C.; Jerez, S. An evaluation of offshore wind power production by floatable systems: A case study from SW Portugal. *Energy* **2017**, *131*, 239–250. [CrossRef]

13. Kim, H.C.; Kim, M.H.; Lee, J.Y.; Kim, E.S.; Zhang, Z. Global Performance Analysis of 5MW WindFloat and OC4 Semi-Submersible Floating Offshore Wind Turbines (FOWT) by Numerical Simulations. In Proceedings of the 27th International Ocean and Polar Engineering Conference, San Francisco, CA, USA, 25–30 June 2017.

14. Kumar, V.; Shrivastava, R.L.; Untawale, S.P. Solar Energy: Review of Potential Green & Clean Energy for Coastal and Offshore Applications. *Aquat. Procedia* **2015**, *4*, 473–480. [CrossRef]

15. Santafé, M.R.; Ferrer Gisbert, P.S.; Sánchez Romero, F.J.; Torregrosa Soler, J.B.; Ferrán Gozálvez, J.J.; Ferrer Gisbert, C.M. Implementation of a photovoltaic floating cover for irrigation reservoirs. *J. Clean. Prod.* **2014**, *66*, 568–570. [CrossRef]

16. Skoplaki, E.; Palyvos, J.A. On the temperature dependence of photovoltaic module electrical performance: A review of efficiency/power correlations. *Sol. Energy* **2009**, *83*, 614–624. [CrossRef]

17. Kamuyu, W.C.L.; Lim, J.R.; Won, C.S.; Ahn, H.K. Prediction model of photovoltaic module temperature for power performance of floating PVs. *Energies* **2018**, *11*, 447. [CrossRef]

18. Trapani, K.; Millar, D.L.; Smith, H.C.M. Novel offshore application of photovoltaics in comparison to conventional marine renewable energy technologies. *Renew. Energy* **2013**, *50*, 879–888. [CrossRef]

19. Wu, Y.; Li, L.; Song, Z.; Lin, X. Risk assessment on offshore photovoltaic power generation projects in China based on a fuzzy analysis framework. *J. Clean. Prod.* **2019**, *215*, 46–62. [CrossRef]

20. Solanki, C.; Nagababu, G.; Kachhwaha, S.S. Assessment of offshore solar energy along the coast of India. *Energy Procedia* **2017**, *138*, 530–535. [CrossRef]

21. Trapani, K.; Millar, D.L. Proposing offshore photovoltaic (PV) technology to the energy mix of the Maltese islands. *Energy Convers. Manag.* **2013**, *67*, 18–26. [CrossRef]

22. Friel, D.; Karimirad, M.; Whittaker, T.; Doran, W.J.; Howlin, E. A review of floating photovoltaic design concepts and installed variations. In Proceedings of the 4th International Conference Offshore Renew Energy CORE 2019, Glasgow, UK, 30 August 2019; pp. 1–10.

23. Pérez-Collazo, C.; Greaves, D.; Iglesias, G.; Perez-Collazo, C.; Greaves, D.; Iglesias, G. A review of combined wave and offshore wind energy. *Renew. Sustain. Energy Rev.* **2015**, *42*, 141–153. [CrossRef]

24. Astariz, S.; Perez-Collazo, C.; Abanades, J.; Iglesias, G. Towards the optimal design of a co-located wind-wave farm. *Energy* **2015**, *84*, 15–24. [CrossRef]

25. Astariz, S.; Perez-Collazo, C.; Abanades, J.; Iglesias, G. Co-located wind-wave farm synergies (Operation & Maintenance): A case study. *Energy Convers. Manag.* **2015**, *91*, 63–75. [CrossRef]

26. Zheng, X.; Zheng, H.; Lei, Y.; Li, Y.; Li, W. An offshore floating wind–solar–aquaculture system: Concept design and extreme response in survival conditions. *Energies* **2020**, *13*, 604. [CrossRef]

27. Iglesias, G.; Carballo, R. Offshore and inshore wave energy assessment: Asturias (N Spain). *Energy* **2010**, *35*, 1964–1972. [CrossRef]

28. Abanades, J.; Flor-Blanco, G.; Flor, G.; Iglesias, G. Dual wave farms for energy production and coastal protection. *Ocean. Coast. Manag.* **2018**, *160*, 18–29. [CrossRef]

29. Ramos, V.; Carballo, R.; Álvarez, M.; Sánchez, M.; Iglesias, G. Assessment of the impacts of tidal stream energy through high-resolution numerical modeling. *Energy* **2013**, *61*, 541–554. [CrossRef]

30. Elliott, D.L.; Holladay, C.G.; Barchet, W.R.; Foote, H.P.; Sandusky, W.F. Wind energy resource atlas of the United States. *STIN* **1987**, *87*, 24819.

31. Zheng, C.W.; Pan, J. Assessment of the global ocean wind energy resource. *Renew. Sustain. Energy Rev.* **2014**, *33*, 382–391. [CrossRef]

32. Prieto, J.I.; Martínez García, J.C.; García, D.; Santoro, R. Notes on the solar map of Asturias. *Renew. Energy Power Qual. J.* **2011**, 1273–1277. [CrossRef]

33. Vázquez, M.V.; Varela, M.P.; Belmonte, P.I.; Cerqueira, M.T.P. Zonas climáticas de radiación solar de Galicia. In Proceedings of the Construyendo el Futuro Sostenible: Libro de Actas del XIV Congreso Ibérico y IX Congreso Iberoamericano de Energía Solar, Vigo, Galicia, España, 17–21 June 2008; pp. 255–260.

34. Samuelsson, P.; Jones, C.G.; Will' En, U.; Ullerstig, A.; Gollvik, S.; Hansson, U.L.; Jansson, E.; Kjellstro, M.C.; Nikulin, G.; Wyser, K. The Rossby Centre Regional Climate model RCA3: Model description and performance. *Tellus A Dyn. Meteorol. Oceanogr.* **2011**, *63*, 4–23. [CrossRef]

35. Uppala, S.M.; Kållberg, P.W.; Simmons, A.J.; Andrae, U.; Bechtold, V.D.C.; Fiorino, M.; Gibson, J.K.; Haseler, J.; Hernandez, A.; Kelly, G.A.; et al. The ERA-40 re-analysis. *Q. J. R. Meteorol. Soc.* **2005**, *131*, 2961–3012. [CrossRef]

36. Cats, G.; Wolters, L. The Hirlam project. *IEEE Comput. Sci. Eng.* **1996**, *3*, 4–7. [CrossRef]

37. Iglesias, G.; Carballo, R. Wave energy and nearshore hot spots: The case of the SE Bay of Biscay. *Renew. Energy* **2010**, *35*, 2490–2500. [CrossRef]

38. Stackhouse, P.W.; Zhang, T.; Westberg, D.; Barnett, A.J.; Bristow, T.; Macpherson, B.; Hoell, J.M. *POWER Release 8 (with GIS Applications) Methodology (Data Parameters, Sources, & Validation) Documentation Date (All Previous Versions Are Obsolete) (Data Version 8.0.1)*; NASA: Washington, DC, USA, 2018.

39. Zhang, T.; Stackhouse, P.W.; Chandler, W.S.; Hoell, J.M.; Westberg, D.; Whitlock, C.H. A global perspective on renewable energy resources: NASA's Prediction of Worldwide Energy Resources (POWER) Project. In Proceedings of the 2007 ISES Solar World Congress, Beijing, China, 18–21 September 2007; Volume 4, pp. 2636–2640.

40. Gelaro, R.; Mccarty, W.; Suárez, M.J.; Todling, R.; Molod, A.; Takacs, L.; Randles, C.; Darmenov, A.; Bosilovich, M.G.; Reichle, R.H.; et al. The modern-era retrospective analysis for research and applications, version 2 (MERRA-2). *J. Clim.* **2017**, *30*, 5419–5454. [CrossRef]

41. Rienecker, M.M.; Suarez, M.J.; Todling, R.; Bacmeister, J.; Takacs, L.; Liu, H.C.; Gu, W.; Sienkiewicz, M.; Koster, R.D.; Gelaro, R.; et al. *The GEOS-5 Data Assimilation System: Documentation of Versions 5.0.1, 5.1.0, and 5.2.0*; NASA Goddard Space Flight Center: Greenbelt, MD, USA, 2008.

42. Oh, K.Y.; Nam, W.; Ryu, M.S.; Kim, J.Y.; Epureanu, B.I. A review of foundations of offshore wind energy convertors: Current status and future perspectives. *Renew. Sustain. Energy Rev.* **2018**, *88*, 16–36. [CrossRef]

43. EWEA European Wind Energy Association: Deep water—The next step for offshore wind energy. 2013.

44. Carrillo, C.; Obando Montaño, A.F.; Cidrás, J.; Díaz-Dorado, E. Review of power curve modelling for windturbines. *Renew. Sustain. Energy Rev.* **2013**, *21*, 572–581. [CrossRef]

45. Ranjbaran, P.; Yousefi, H.; Gharehpetian, G.B.; Astaraei, F.R. A review on floating photovoltaic (FPV) power generation units. *Renew. Sustain. Energy Rev.* **2019**, *110*, 332–347. [CrossRef]

46. Patil, S.S.; Wagh, M.M.; Shinde, N.N. A Review on Floating Solar Photovoltaic Power Plants. *Int. J. Sci. Eng. Res.* **2017**, *8*, 789–794.

47. Trapani, K.; Millar, D.L. The thin film flexible floating PV (T3F-PV) array: The concept and development of the prototype. *Renew. Energy* **2014**, *71*, 43–50. [CrossRef]

48. Parida, B.; Iniyan, S.S.; Goic, R. A review of solar photovoltaic technologies. *Renew. Sustain. Energy Rev.* **2011**, *15*, 1625–1636. [CrossRef]

49. Jinkosolar. Eagle 72P-V 320-340 Watt. 2017. Available online: https://jinkosolar.eu (accessed on 15 June 2019).

50. Trinasolar. Tallmax Plus TSM-DE14A (II). 2017. Available online: https://www.trinasolar.com (accessed on 15 June 2019).

51. Q-CELLS. Q.POWER-G5 260-280. 2017. Available online: https://www.q-cells.com (accessed on 15 June 2019).

52. Del Jesus, F.; Menéndez, M.; Guanche, R.; Losada, I.J. A wind chart to characterize potential offshore wind energy sites. *Comput. Geosci.* **2014**, *71*, 62–72. [CrossRef]

53. Ramos, V.; Iglesias, G. Wind power viability on a small island. *Int. J. Green Energy* **2014**, *11*, 741–760. [CrossRef]

54. Wind Europe. *Offshore Wind in Europe. Key Trends and Statistics 2017*; Wind Europe: Brussels, Belgium, 2018.

55. Weaver, T. Financial appraisal of operational offshore wind energy projects. *Renew. Sustain. Energy Rev.* **2012**, *16*, 5110–5120. [CrossRef]

56. Sukarso, A.P.; Kim, K.N. Cooling effect on the floating solar PV: Performance and economic analysis on the case of west Java province in Indonesia. *Energies* **2020**, *13*, 2126. [CrossRef]

57. Gao, X.; Yang, H.; Lu, L. Investigation into the optimal wind turbine layout patterns for a Hong Kong offshore wind farm. *Energy* **2014**, *73*, 430–442. [CrossRef]

58. Hundleby, G.; Freeman, K. *Unleashing Europe's Offshore Wind Potential: A New Resource Assessment*; BVG Associates: Swindon, UK, 2017.

59. Astariz, S.; Iglesias, G. Output power smoothing and reduced downtime period by combined wind and wave energy farms. *Energy* **2016**, *97*, 69–81. [CrossRef]

60. Carballo, R.; Sanchez, M.; Ramos, V.; Fraguela, J.A.; Iglesias, G. The intra-annual variability in the performance of wave energy converters: A comparative study in N Galicia (Spain). *Energy* **2015**, *82*, 138–146. [CrossRef]

Journal of
*Marine Science
and Engineering*

Article

Assessing the Effectiveness of a Novel WEC Concept as a Co-Located Solution for Offshore Wind Farms

Victor Ramos [1,2,*], Gianmaria Giannini [1,2], Tomás Calheiros-Cabral [1,2], Mario López [3], Paulo Rosa-Santos [1,2] and Francisco Taveira-Pinto [1,2]

1   Department of Civil Engineering, Faculty of Engineering of the University of Porto (FEUP),
    Rua Dr. Roberto Frias, S/N, 4200-465 Porto, Portugal; gianmaria@fe.up.pt (G.G.); tcabral@fe.up.pt (T.C.-C.);
    pjrsantos@fe.up.pt (P.R.-S.); fpinto@fe.up.pt (F.T.-P.)
2   Interdisciplinary Centre of Marine and Environmental Research of the University of Porto (CIIMAR),
    Avenida General Norton de Matos, S/N, 4450-208 Matosinhos, Portugal
3   DyMAST Research Group, Department of Construction and Manufacturing Engineering, University of
    Oviedo, EPM, C/Gonzalo Gutiérrez Quirós S/N, 33600 Mieres, Asturias, Spain; mario.lopez@uniovi.es
*   Correspondence: josevictor.ramos@fe.up.pt

Abstract: The combined exploitation of wave and offshore wind energy resources is expected to improve the cost competitiveness of the wave energy industry as a result of shared capital and operational costs. In this context, the objective of this work is to explore the potential benefits of co-locating CECO, an innovative wave energy converter, with the commercial *WindFloat Atlantic* wind farm, located on the northern coast of Portugal. For this purpose, the performance of the combined farm was assessed in terms of energy production, power smoothing and levelised cost of energy (LCoE). Overall, the co-located farm would increase the annual energy production by approximately 19% in comparison with the stand-alone wind farm. However, the benefits in terms of power output smoothing would be negligible due to the strong seasonal behaviour of the wave resource in the area of study. Finally, the LCoE of the co-located farm would be drastically reduced in comparison with the stand-alone wave farm, presenting a value of 0.115 per USD/kWh, which is similar to the levels of the offshore wind industry as of five years ago. Consequently, it becomes apparent that CECO could progress more rapidly towards commercialisation when combined with offshore wind farms.

Keywords: marine renewable energy; offshore wind energy; wave energy; CECO; *WindFloat Atlantic*; co-located wind–wave farm

Citation: Ramos, V.; Giannini, G.; Calheiros-Cabral, T.; López, M.; Rosa-Santos, P.; Taveira-Pinto, F. Assessing the Effectiveness of a Novel WEC Concept as a Co-Located Solution for Offshore Wind Farms. *J. Mar. Sci. Eng.* **2022**, *10*, 267. https://doi.org/10.3390/jmse10020267

Academic Editors: Constantine Michailides and Domenico Curto

Received: 16 December 2021
Accepted: 10 February 2022
Published: 15 February 2022

Publisher's Note: MDPI stays neutral with regard to jurisdictional claims in published maps and institutional affiliations.

1. Introduction

In recent years, concerns related to climate change and energetic sustainability have prompted different policies (e.g., *Paris Agreement* [1], 2030 Agenda for Sustainable Development [2] and EU's *Green Deal* [3]) to foment a transition towards a sustainable and carbon-neutral economy. Within the emerging renewable energy sources, marine renewable energy (MRE), which presents a vast, geographically diverse and virtually untapped resource (up to 32 TW [4]), is expected to play a crucial part in achieving the above-mentioned goals. Previous studies have analysed the potential benefits of developing a fully fledged MRE industry [5]. First of all, MRE could contribute to increasing and diversifying the current low-carbon generation portfolio (in the order of 330 and 550 GW by 2050 [5]). Second, MRE could supply a significant share of future energy demands (e.g., up to 10% of EU's energy needs by 2050 [5]). Finally, MRE appears as a fantastic opportunity to develop a new industrial sector and boost the economy of coastal regions, contributing to their long-term sustainability [5].

Among the large variety of MRE sources, offshore wind and wave energy have attracted greater interest from the academic and industrial communities [6]. On the one hand, offshore wind energy, with an estimated global resource of 71,000 GW, has experienced substantial growth over the last decade. In 2020, offshore wind capacity has exceeded

35 GW, representing 4.8% of the global wind capacity [7]. In this context, the EU appears as a global front-runner, presenting an installed capacity of 24.92 GW [8] and future goals to reach between 230 and 450 GW by 2050 [9]. Several factors, including developer experience, standardisation of turbine foundations, improvements in wind turbine technology, better practices during installation and maintenance operations, and economies of scale have all contributed to this rapid development [10]. As result, in the period between 2010 and 2020, the levelised cost of energy (LCoE) has dropped by 48%, from 0.162 to 0.084 USD/kWh [10]. However, significant progress is still required to make offshore wind energy cost-competitive with other renewable sources, such as onshore wind or solar PV, which present LCoE values of 0.039 and 0.059 USD/kWh, respectively [10]. For this purpose, the offshore wind industry is trending towards deeper water locations, which present stronger wind conditions [7], and floating turbines with higher hub heights and longer blades, which present larger power capacity [11].

On the other hand, wave energy, despite presenting a large and globally diverse resource [12], is still far from reaching the commercialisation stage [13]. This fact is intimately connected to the lack of progress of wave energy converter (WEC) technology. Over the last two decades, a multitude of WEC concepts, based on different working principles, has been put forward. Among them, point absorbers (e.g., CorPower [14,15]), overtopping devices (e.g., Wavedragon [16] and Wavecat [17]), oscillating water columns [18,19], attenuators (e.g., Pelamis [20]), as well as oscillating wave surge (e.g., Oyster [21]) and submerged differential pressure (e.g., CETO [22]) devices, stand out. However, the great majority of them are still present at low technology readiness levels, far from commercial viability [23]. As a result, LCoE values for wave energy are approximately ten times higher than other renewable sources such as onshore wind and solar PV [24]. In consequence, significant effort is still required to increase the cost competitiveness of wave energy. In this context, improving WEC performance and reliability becomes essential to decrease the associated LCoE. In addition, the negligible development of WEC technology has inevitably led to a lack of industrial expertise in terms of supply chain, logistics and operational tasks (including WEC deployment, maintenance, grid integration and decommissioning) [25].

On these grounds, it becomes apparent that offshore wind and wave energies present certain similarities in terms of exploitation locations, conversion technologies, operation and maintenance tasks, grid integration and logistics [26]. Consequently, potential synergies between the two energy forms could contribute to improving significantly their cost competitiveness [27]. Among the different methodologies available to exploit simultaneously offshore wind and wave energy resources, co-located wind–wave arrays have emerged as the most feasible option [6]. This approach offers valuable mutual benefits for wind and wave operations [28]. First, a huge opportunity for shared costs arises [29]. Concerning capital costs, wind and wave operations could share expenses in terms of sea space leasing, consenting procedures, electrical infrastructure (export cable, offshore and onshore substations) and onshore facilities [30]. In addition, operational costs can also be reduced by sharing logistics and integrating operation and maintenance tasks [31]. In this context, Astariz et al. observed that, for the German and Danish coasts, co-located wave energy farms presented LCoE reductions ranging between 55 and 70%, with respect to stand-alone wave farms [32,33]. Another benefit of combining wind and wave exploitation is the reduction in power output variability (i.e., power smoothing) of offshore wind farms [34]. This fact is supported by the higher predictability of wave conditions and the lag between peak wind speeds and peak wave heights (especially in regions dominated by swell waves) [35,36]. In this sense, Astariz et al. found reductions in downtime and power variability for the North Sea *Alpha Ventus* and *Horns Rev* offshore wind farms of approximately 87 and 6%, respectively, due to the presence of co-located wave farms [37]. Similarly, Gaughan and Fitzgerald found that, in the Irish West coast, the power variability of offshore wind farms could be significantly reduced by the co-location of wave energy farms [38]. On the other hand, the operation of a co-located WEC array may result in significant wave height reductions in its lee (known as the shadow effect [33,39]), facilitating accessibility and maintenance operations of wind farms [40]. For instance, Astariz et al. found that the sheltering effect of co-located wave farms could increase up to 20% the annual accessibility to offshore wind farms located in the North Sea [33,41]. Last but not least, a transfer of knowledge, including industrial best practices and operational standards, could flow from the wind to the wave energy sectors.

In consequence, certain WEC technologies (mainly offshore floating devices) could progress more rapidly towards commercialisation if they were used as a co-located solution for offshore wind farms, benefiting from the offshore wind industry as result of cost sharing and transfer of knowledge. Among the different WEC technologies that could benefit from the synergies with the offshore wind industry is the CECO device [42]. CECO is a novel WEC concept that simultaneously harnesses the kinetic and the potential energy of the waves by means of oblique motion (Section 2.2.1). As a result of this innovative configuration, CECO has shown a promising potential for harnessing wave energy [43]. Furthermore, CECO presents a broad range of operating water depths, making it suitable for co-location with offshore wind farms. Against the foregoing backdrop, the objective of this paper is to assess the effectiveness and potential benefits of co-locating the CECO device with a commercial floating offshore wind farm. In this context, the performance of CECO as a co-located solution was assessed in terms of energy output, power smoothing and LCoE reductions. To date and to the best knowledge of the authors, no previous research has addressed these issues in detail. For this purpose, the northern coast of Portugal, where the *WindFloat Atlantic* offshore wind farm is located [44], was used as case study (Figure 1). This region, facing the North Atlantic, presents one of the most energetic wind and wave regimes in continental Europe [45,46]; therefore, it appears as a promising location for exploiting both the offshore wind and wave energy resources.

The remainder of this paper is structured as follows: Section 2 describes the area of study and the characteristics of wave (CECO) and wind conversion technologies and presents the methodology used to evaluate the effectiveness of the proposed co-located wind–wave array. Section 3 presents the results obtained. Section 4 discusses the main advantages and disadvantages of the proposed co-located wind–wave array. Finally, conclusions are drawn in Section 5.

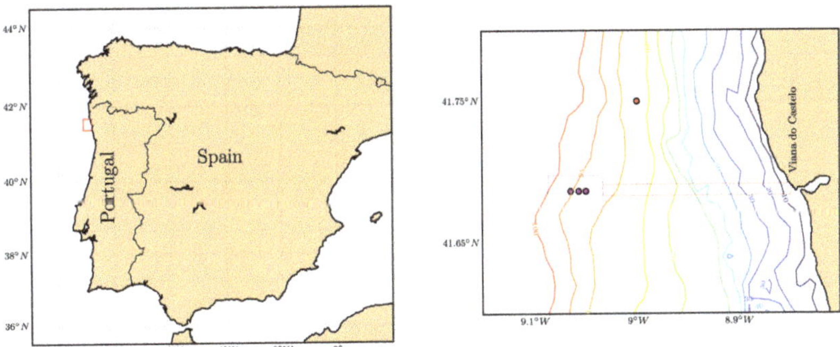

Figure 1. Location of the area of study. (Colour lines represent the bathymetric isolines. Red and magenta dots highlight the position of the SIMAR dataset and wind turbines, respectively. Blue dashed line represents the allocated sea space for the *WindFloat Atlantic* farm. Red dashed line represents submarine cables area).

2. Materials and Methods

2.1. Area of Study

For the present work, the *WindFloat Atlantic* offshore wind farm was used as a case study [44]. Located off the northern coast of Portugal (Figure 1), the water depth and distance to shore are in the order of 100 m and 15 km, respectively. *WindFloat Atlantic* is composed of three turbines (of 8.4 MW each) mounted on semi-submersible floating platforms anchored to the seabed. The turbines are aligned in a west–east direction, separated by approximately 600 m. It is worth pointing out that the allocated sea surface for the wind farm spans an area of 11.25 km^2 (Figure 1), opening the possibility for future wind farm expansions or co-location of WEC arrays. Finally, *WindFloat Atlantic* has available excellent port infrastructures. For the day-to-day operation and maintenance tasks, a base was set up in the the Port of Viana do Castelo, which is located at a 20 km distance. For manufacturing and assembly purposes, the outer Port of Ferrol (Spain), located approximately at 270 km distance, appears as an excellent alternative [47].

Besides its outstanding wind energy resource [45], the northern coast of Portugal also presents a very energetic wave climate, with mean annual values of wave power (per meter of wave front) in the order of 25 kWm^{-1} [48]. Nonetheless, the wave regime of the region also presents a strong inter- and intra-annual variability [46,48]. This behaviour is governed by the North Atlantic Oscillation (NAO), which corresponds to a large scale meridional oscillation of atmospheric mass between the Azores High and the Iceland Low. Its behaviour is defined by the so-called NAO index that can be positive or negative, corresponding to a high or low pressure difference between the Azores High and the Iceland Low, respectively. Therefore, periods with positive NAO values result in strong wave regimes with a predominant north–west direction. Conversely, periods with negative NAO values derive into weaker wave regimes shifted towards a more westerly direction. As a result, the mean seasonal values of wave power range from summer to winter conditions from 10 kWm^{-1} to 45 kWm^{-1}, respectively [46,48]. In consequence, this area appears as a perfect location for the installation of co-located wave-wind arrays, therefore, for evaluating their potential benefits and synergies under a wide range of wave conditions.

2.2. Co-Located Wave–Wind Farm

2.2.1. Energy Conversion Technologies

This section presents the wave and wind conversion technologies considered for the present work. With the aim of capturing wave energy, the CECO device was used as case study. CECO is a (oscillating body) point absorber WEC, whose main novelty resides in presenting a sloped power take-off (PTO) system [49]. This PTO configuration allows it to capture both the vertical and horizontal force components of ocean waves [50,51]. In consequence, wave energy is absorbed by a floating body with its motions restrained to an inclined direction. The main elements of the floating body are two lateral mobile modules (LMMs) joined by a frame of tubular elements. In its current version, CECO uses a rack pinion system to transform the absorbed energy into electricity (Figure 2). Therefore, the pinion is housed in the interior of the supporting element, while the rack is mounted on the floating frame and oscillates with it, driving a rotatory electric generator. Nonetheless, for future designs, a gearbox linked to the pinion or a linear electric generator may be adopted to reduce the energy losses in the transmission and generation stages of CECO [43].

Over the last five years, extensive research has been conducted to refine the original CECO design. First, the shape and mass of the LMMs have been optimised to improve the hydrodynamic response of CECO [52,53] and, consequently, increase the power absorption for a wider range of wave conditions [42]. Furthermore, it was found that both the PTO damping [54] and inclination [55] also play a predominant role in CECO's response. In consequence, relevant variables related to the LMMs (mass and submergence levels) and PTO system (damping and inclination) could be adjusted (by a control system) to match the resonance condition of the most frequent sea states, thus improving CECO's power absorption.

In addition, CECO offers high flexibility in terms of installation. For nearshore locations (<30 m of water depth), the original CECO design was based on a fixed-bottom support structure (monopile, tripod or jacket foundations). However, for deeper locations, the use of a fixed support structure becomes economically unfeasible; therefore, a floating support structure is required. In this context, the floating version of CECO was designed and optimised for the wave conditions of the Portuguese coast, using an operational-limit sea state of $H_s = 7.5$ m and $Tp = 16$ s, with a configuration based on a tension leg platform with a star-shape mooring system offering the best results in terms of energy production and mooring loads (Figure 2) [56]. For the present work, the floating version of CECO defined by Giannini et al. [56] was used as case study. The main design characteristics of the device are summarised in Table 1, while its (absorption) power matrix is shown in Figure 3.

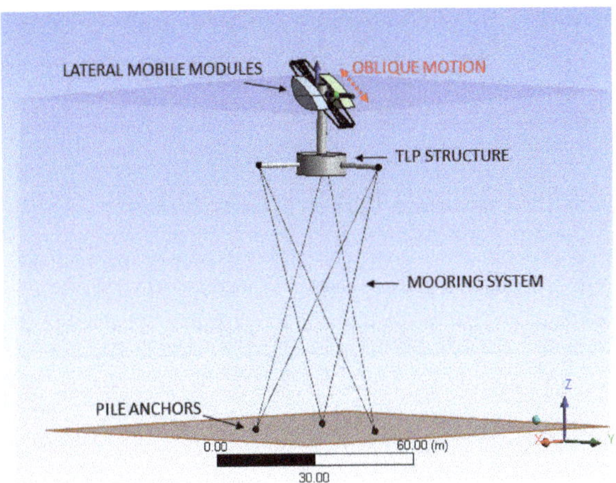

Figure 2. Main parts of the CECO device for offshore installations.

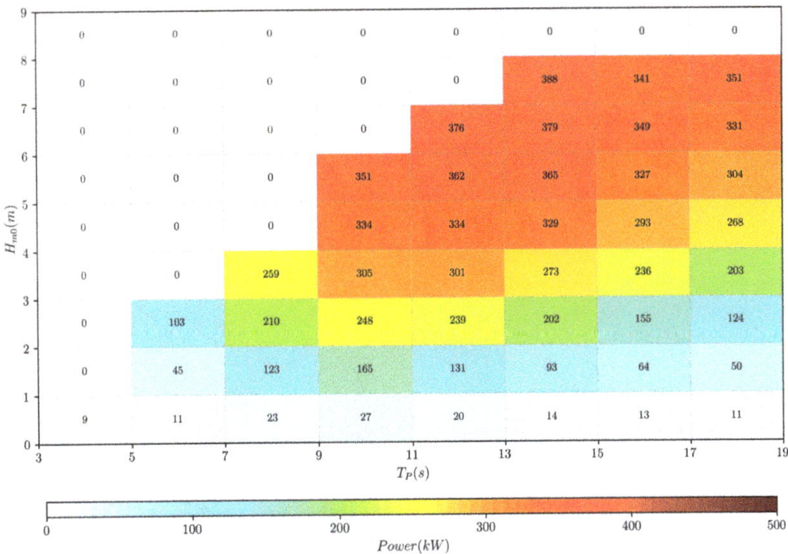

Figure 3. Power matrix of CECO device. (Numbers represent the mean power production for each sea-state bin. Colourmap represents the power produced for the different sea states).

Table 1. Main design characteristics of CECO wave energy converter.

Parameter	Value
PTO inclination angle (°)	30
LMM inclination (°)	45
LMM length (m)	9.52
LMM width (m)	6
LMM maximum stroke (m)	15
LMM mass (ton)	288
Overall width (m)	22
PTO rated power (MW)	0.5

Regarding wind conversion technology, *WindFloat Atlantic* uses the *MHI-Vestas V164-8.3 MW* offshore wind turbine, which, to date, appears as the most powerful turbine used for a floating wind farm [57]. As indicated in Section 2.1, the turbines are mounted on an innovative three-column semi-submersible floating structure (*WindFloat* ® [58]), which is anchored to the seabed. The floating structure achieves stability by combining the use of damping plates with a static and dynamic ballast system. As a result, the mounting structure is able to withstand wave heights and wind speeds exceeding 17 m and 100 kmh^{-1}, respectively [58,59]. Due to the lack of detailed technical information, for the present work, the *Vestas V164-8.0 MW* wind turbine was used as case study, since its characteristics are very similar to the *MHI-Vestas V164-8.3 MW*. The main technical specifications and power curve of the wind turbine are shown in Table 2 and Figure 4, respectively [60].

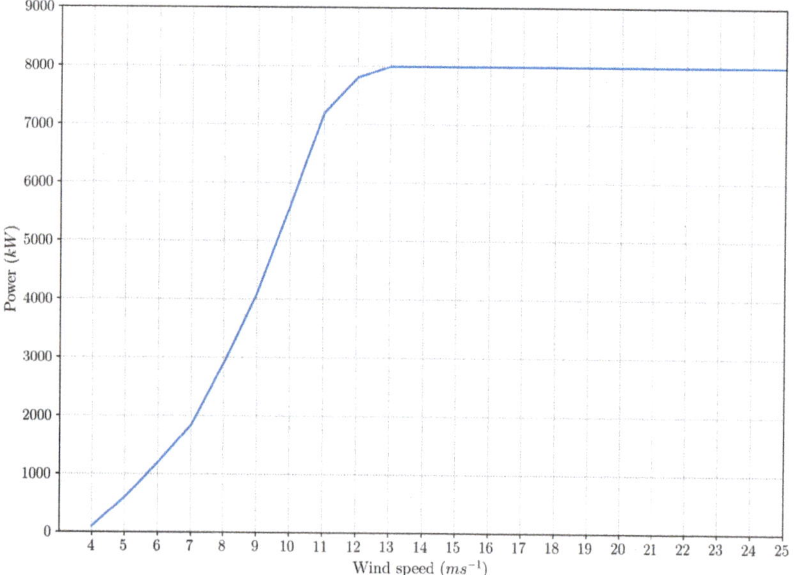

Figure 4. Power curve of *Vestas V164-8.0 MW* wind turbine.

Table 2. Main technical specification of the *Vestas V164-8.0 MW* wind turbine.

Parameter	Value
Rated power (MW)	8.0
Rotor diameter (m)	164
Hub height (m)	105
Cut-in wind speed (ms^{-1})	4
Rated wind speed (ms^{-1})	13
Cut-off wind speed (ms^{-1})	25
Survival wind speed (ms^{-1})	50

Based on Figure 4, the power curve of the *Vestas V164-8.0 MW* turbine can be expressed analytically as follows:

$$P(V) = \begin{cases} 0, & V < V_I \\ (a_n V^n + a_{n-1} V^{n-1} + \dots + a_2 V^2 + a_1 V + a_0), & V_I \leq V < V_R \\ P_R, & V_R \leq V < V_0 \\ 0, & V \geq V_0 \end{cases} \tag{1}$$

where $a_n, a_{n-1}, \dots, a_2, a_1, a_0$ are the polynomial coefficients of the power curve, while V_R and P_R are, respectively, the rated velocity and rated power of the wind turbine (Table 2). In this case, the values of the polynomial coefficients, obtained from a sixth order polynomial regression fit, are summarised in Table 3.

Table 3. Polynomial coefficients of the *Vestas V164-8.0 MW* power curve.

Parameter	Value
a_6 ($m^{-6}s^6$)	1.493×10^{-1}
a_5 ($m^{-5}s^5$)	-7.358
a_4 ($m^{-4}s^4$)	1.418×10^2
a_3 ($m^{-3}s^3$)	-1.367×10^3
a_2 ($m^{-2}s^2$)	7.028×10^3
a_1 ($m^{-1}s$)	-1.794×10^4
a_0 ($-$)	1.751×10^4

2.2.2. Layout Design

In this section, a preliminary design for the co-located CECO array is proposed. The array, with a total capacity of 5 MW, is formed by a single row of devices, which follows a curvilinear trajectory oriented to the prevailing NW wave direction (Figure 5). Due to the lack of previous research dealing with the CECO park effects, a conservative separation of 330 m (approximately 15 times the overall width of CECO) between devices was chosen. The distance between the wave and wind arrays was set to be higher than 800 m, with the aim of avoiding potential disruptions during operational and maintenance tasks. As a result, it was assumed that wave and wind arrays do not share mooring or structural elements. In addition, the proposed design ensures that all CECO devices lie inside the allocated sea space for the *WindFloat Atlantic* wind farm, saving extra cost in terms of sea-space consenting and leasing. It is worth mentioning that the optimal design of the array, in terms of energy output and shadow effect [61], is beyond the scope of this work; therefore, the array was designed to assess the feasibility of CECO as a co-located solution. Finally, the main characteristics of the co-located wave–wind farm are summarised in Table 4.

Table 4. Main characteristics of the co-located wind–wave farm.

Parameter	Value
Number of WECs	10
Number of wind turbines	3
Rated power of WEC array (MW)	5
Rated power of wind turbine array (MW)	24
Rated power of co-located farm (MW)	29
Spacing between WECs (m)	330
Spacing between wind turbines (m)	600
Water depth (m)	95

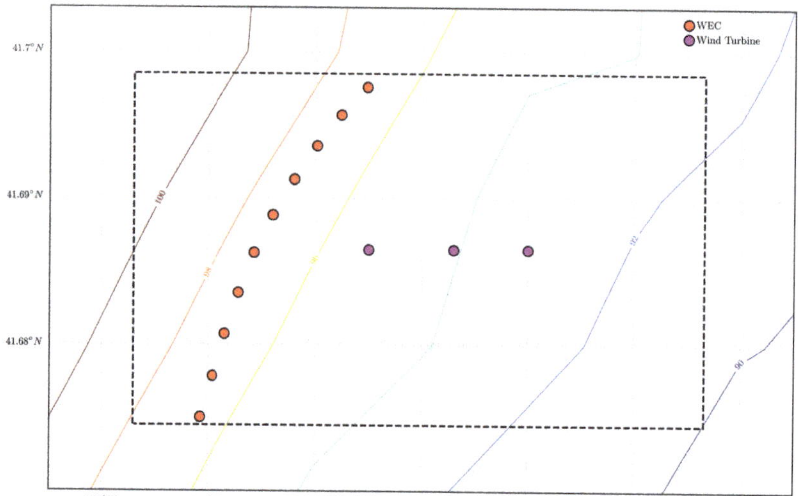

Figure 5. Spatial distribution of the proposed co-located wave-wind farm. (Colour lines represent the bathymetric isolines. Red and magenta dots highlight the position of WECs and wind turbines, respectively. Black dashed line represents the allocated sea space for the *WindFloat Atlantic* farm).

2.3. Performance Assessment of the Co-Located Wave-Wind Farm

This section presents the methodology used for assessing the performance of the co-located wave–wind farm proposed in Section 2.2. The methodology focuses on evaluating the energy output of the co-located wind–wave farm in comparison with the individual wind and wave farms (Section 2.3.1), the reduction in power production variability (Section 2.3.2) and the potential LCoE reductions (Section 2.3.3).

2.3.1. Energy Output

The energy output produced by the co-located wind–wave farm (E_{wwf}) can be expressed as

$$E_{wwf} = E_{wndf} + E_{wavf},\tag{2}$$

where E_{wndf} and E_{wavf} are the energy outputs produced by the wind and wave farms, respectively. In this sense, the energy production of the wind farm is calculated as follows:

$$E_{wndf} = n_{wt}E_{wt},\tag{3}$$

where n_{wt} is the total number of wind turbines and E_{wt} is the energy output produced by an individual wind turbine, which can be computed as

$$E_{wt} = T \int_{V_i}^{V_0} P(V)f(V)dV, \tag{4}$$

where T is the period of time considered, V_i and V_0 are, respectively, the cut-in and cut-off velocity values of the turbine, $P(V)$ is the power curve of the wind turbine (Equation (1)) and V the wind velocity at the hub height of the wind turbine. Assuming a logarithmic wind velocity profile, V can be obtained as

$$V = V_{z_0} \left(\frac{z}{z_0} \right)^{\alpha}, \tag{5}$$

where z is the hub height, V_{z_0} is the wind speed at a reference height z_0 (in this case, a 10 m height) and α is an empirical coefficient that depends on the surface roughness characteristics. Based on previous studies on the region, a value of $\alpha = 0.056$ was used [62]. Finally, $f(V)$ is the wind probability distribution. For the present work, the well-known Weibull distribution was used. Therefore, $f(V)$ can be expressed as

$$f(V) = \frac{k}{c} \left(\frac{V}{c} \right)^{(k-1)} e^{-\left(\frac{V}{c}\right)^k}, \tag{6}$$

where k and c are the so-called scale and shape parameters of the Weibull distribution, respectively. In consequence, combining Equations (1), (4) and (6), E_{wt} can be expressed as

$$\begin{aligned} E_{wt} = T \Bigg\{ &\int_{V_i}^{V_R} \left(a_n V^n + a_{n-1} V^{n-1} + \ldots + a_1 V + a_0 \right) \frac{k}{c} \left(\frac{V}{c} \right)^{(k-1)} e^{-\left(\frac{V}{c}\right)^k} dV \\ &+ \int_{V_R}^{V_0} P_R \frac{k}{c} \left(\frac{V}{c} \right)^{(k-1)} e^{-\left(\frac{V}{c}\right)^k} dV \Bigg\}, \end{aligned} \tag{7}$$

On the other hand, the energy output produced by the wave farm E_{wavf} can be expressed as

$$E_{wavf} = n_{wec} E_{wec}, \tag{8}$$

where n_{wec} is the total number of WECs forming the wave farm and E_{wec} is the energy output produced by an individual WEC, which can be calculated according to Equation (9).

$$E_{wec} = \sum_{i=1}^{n} P_i O_i, \tag{9}$$

where P_i is the power generated by the WEC for the *i-th* sea state (Figure 3) and O_i is the occurrence for a certain reference period (year, month or day) of the i-th sea state.

To compute the energy production of the farms, the wind and wave conditions of the area of study were obtained from the SIMAR datasets. SIMAR is an hourly re-analysis dataset managed by the Spanish Port Authority (Puertos del Estado), which covers the period from 1958 to date. Wind and wave conditions were computed for the North Atlantic and the Mediterranean Sea, with a spatial resolution of 0.25° × 0.25°, by means of joint numerical modelling of atmospheric, sea level and wave conditions [63]. From the SIMAR dataset, hourly wind and wave data, covering the period from 1 January 1960 to 31 December 2020, were obtained for the location highlighted in Figure 1. This large number of data allowed us to estimate the energy outputs on an annual and monthly bases. Finally, due to the simplicity of the proposed co-located wind–wave array (Section 2.2.2), the wave and wind park effects were not considered when computing energy production.

2.3.2. Power Smoothing

The contributions of the CECO array to reduce the power output variability of the *WindFloat Atlantic* farm was assessed by means of the Power Smoothing Index (PSI) [62], which can be computed by means of Equation (10).

$$PSI = \begin{cases} \frac{CV_{wnd} - CV_{wnd-wav}}{CV_{wnd}}, & CV_{wnd} > CV_{wnd-wav} \\ 0, & CV_{wnd} \leq CV_{wnd-wav} \end{cases} \tag{10}$$

where CV_{wnd} and $CV_{wnd-wav}$ are the variation coefficients for the energy outputs of the wind and co-located wind–wave farms, respectively. Overall, PSI values range in a scale from zero to one. Therefore, a PSI value of one indicates that the power variability, for a certain reference period, is eliminated. Conversely, PSI values close to zero indicate that reductions in power output variability are almost negligible.

2.3.3. Cost Benefits

With the aim of assessing the potential cost benefits of the co-located wind–wave farm proposed in Section 2.2.2 (Figure 5), the LCoE was computed and compared for three different scenarios, the stand-alone wind farm (*WindFloat Atlantic*), the stand-alone wave farm (CECO array) and the co-located wind–wave farm. For this purpose, the analytical cost model proposed by Clark et al. [30], which was developed ad hoc for co-located floating wind–wave farms, was used. The model computes the LCoE using a life-cycle cost approach [30].

$$LCoE = \frac{\sum_{t=0}^{T} \frac{C_t}{(1+r)^t}}{\sum_{t=0}^{T} \frac{E_t}{(1+r)^t}}, \tag{11}$$

where T is the lifespan of the project in years; C_t and E_t are the costs and energy output for the t-th year of the project, respectively; and r is the discount rate, which can be computed as

$$r = \frac{r_{inflation} + r_{loan}}{1 - r_{inflation}}, \tag{12}$$

where $r_{inflation}$ and r_{loan} are the inflation and loan rates, respectively. Following the work of Clark et al. [30], for the present study, values of 2% inflation rate and 10% loan rate were used. With respect to the costs, Clark et al. [30] classified them into four main categories, pre-installation (C_{prei}), implementation (C_{imp}), operational (C_{op}) and decommissioning costs (C_{decm}).

$$C_t = C_{prei} + C_{imp} + C_{op} + C_{decm}, \tag{13}$$

Pre-installation costs include expenses related to market analysis, resource and feasibility assessment, site selection, metocean data acquisition, engineering design and consenting and licensing procedures. For this cost item, Clark et al. [30] suggested the use of different values found in the literature. For the present work, the value of USD 250,000 per installed MW, proposed by Myhr et al. [64], was considered.

On the other hand, implementation costs include expenses covering the design (C_{desgin}), construction (C_{const}) and installation (C_{inst}) of conversion technologies and electrical subsystems, such as substations ($C_{substat}$), as well as inter-array and export cables (C_{cable}).

$$C_{imp} = C_{design} + C_{const} + C_{inst} + C_{substat} + C_{cable}, \tag{14}$$

Table 5 summarises the analytical expressions proposed by Clark et al. [30], for each of the sub-costs of Equation (14).

With regards to the operational costs, these include O&M ($C_{O\&M}$), administrative(C_{adm}) and insurance (C_{ins}) expenses.

$$C_{op} = C_{O\&M} + C_{adm} + C_{ins}, \tag{15}$$

Similarly, the values and analytical expressions proposed by Clark et al. [30], for each of the sub-costs of Equation (15), are summarised in Table 6.

Table 5. Values and analytical expressions proposed by Clark et al. [30] for implementation costs of wind and wave energy farms (l_{moor}, length of the anchoring and mooring lines; l_{array}, length of inter-array cable; l_{export}, length of export cable).

Costs (USD)		Wind	Wave
Conversion technology	Design Construction Installation	$C_{design} = 240,000$ $C_{const} = 1096[1350P_R + n_{wt}(511 + l_{moor})]$ $C_{inst} = 977,620(n_{wt} + n_{wec})$	$C_{design} = 240,000$ $C_{const} = 1,519,037 n_{wec}$
Electrical systems	Substation Cable	$C_{substat} = 20,000(100 + P_R)$ $C_{cable} = 307(l_{array} + 1.60 l_{export})$	

Table 6. Values and analytical expressions proposed by Clark et al. [30] for operational costs of wind and wave energy farms (r_{ins}, insurance rate of the project).

Costs (USD)	Wind	Wave	Co-Located Wave–Wind
O&M Administrative Insurance	$C_{O\&M,wt} = 133,000 P_R T$	$C_{O\&M,wec} = 228,564 P_R T$ $C_{adm} = 3,000,000$ $C_{ins} = r_{ins} C_t$	$C_{O\&M} = 0.82(C_{O\&M,wt} + C_{O\&M,wec})$

Finally, decommissioning costs can be computed according to Equation (16).

$$C_{decm} = r_{decm} C_t, \tag{16}$$

where r_{decm} is the decommissioning rate, which, for the present work, was set to 3% [30]. For further details of the cost model, the readers are referred to the work of Clark et al. [30].

3. Results

3.1. Energy Output

The energy output produced by the wind, wave and co-located farms was computed on annual (Section 3.1.1) and monthly (Section 3.1.2) bases, using the methodology presented in Section 2.3.1.

3.1.1. Energy Output: Annual Scenario

For the stand-alone wind farm, the hourly wind velocities at the hub height (105 m) of the wind turbine were computed by means of Equation (5) from the SIMAR reanalysis datasets (Section 2.3.1) and then fitted to the Weibull probability density function, as can be observed in Figure 6. Overall, the wind distribution presents its largest frequency for wind speeds in the order of 7 ms^{-1}, ensuring the operation of the *Vestas V164-8.0 MW* wind turbine for more than 78% of annual hours. Furthermore, the fraction of time for which the turbine operates at its rated power exceeds 9%, representing approximately 630 annual hours. On these grounds, the annual energy production of the wind farm was obtained combining Equations (3) and (7), yielding a value of 70.44 gigawatt hours per year (GWh/a). This value is in good agreement with the annual energy output reported by *WindFloat Atlantic* during its first year of operation [65].

On the other hand, the SIMAR hourly records of wave data (spanning from 1 January 1960 to 31 December 2020) were used to assess the wave energy resource in the area of study. For this purpose, the omni-directional wave energy matrix (Figure 7) was constructed in terms of significant wave height (H_s), peak period (T_P) and annual number of hours of occurrence (O_i). In general, the most recurrent sea states are concentrated in the range of 1–3 m of H_s and 9–13 s of T_P. However, the bulk of wave energy shifts towards the region of 2–5 m of H_s and 9–15 s of T_P, with some sea states exceeding values of 20 MWhm^{-1}. On these grounds, the annual energy production of the wave farm was computed by combining the CECO and wave resource matrices as indicated in Equations (8) and (9). Therefore, the annual energy output of the wave farm resulted as 13.52 GWh/a.

As a result, the co-located wind–wave farm yields a total annual energy production of approximately 84 GWh/a, with the wind and wave farms contributing 84% and

16%, respectively. Finally, Table 7 summarises the values and contributions of the energy production for the annual scenario.

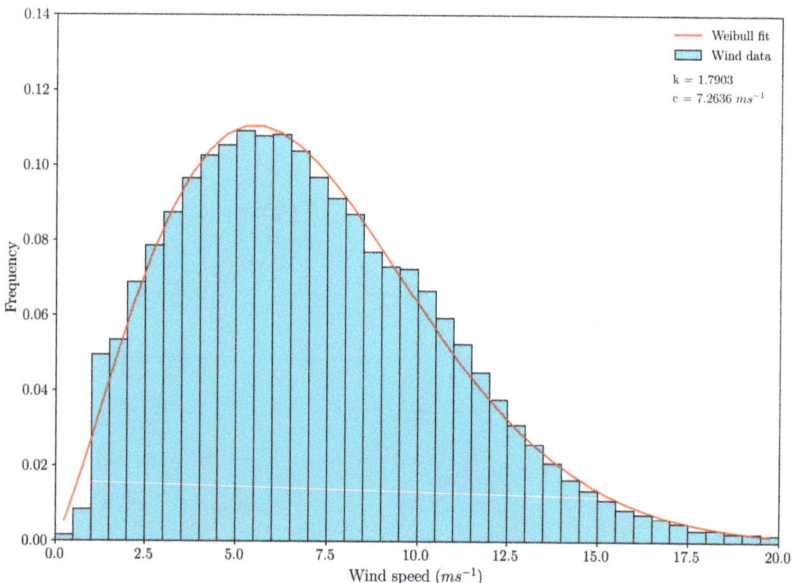

Figure 6. Observed wind data (histogram) and Weibull probability density function (red line) for the annual scenario (k and c are the annual scale and shape parameter values of the Weibull distribution).

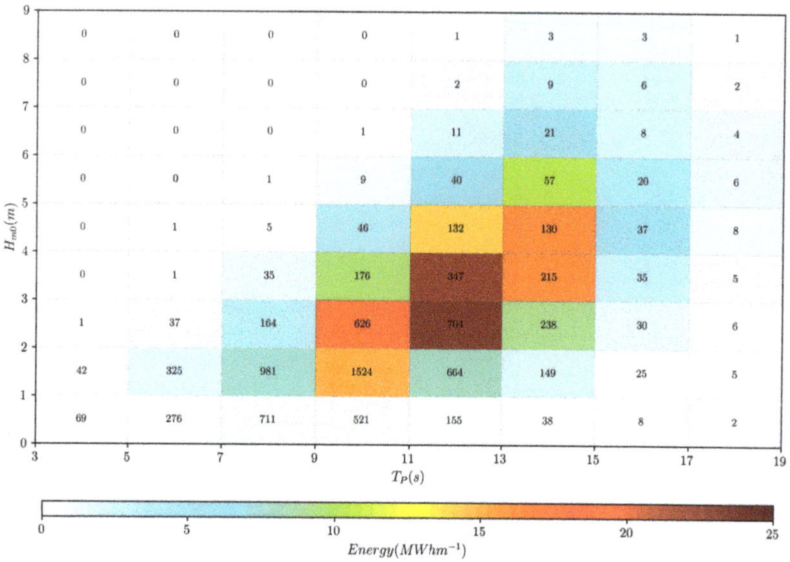

Figure 7. Wave resource energy matrix. (Colourmap represents the annual energy per meter of wave front and the numbers represent the occurrence in annual hours for each sea state (O_i)).

Table 7. Annual energy production for the co-located and stand-alone wind and wave farms.

Parameter	Value
Energy output of single wind turbine (GWh)	23.48
Energy output of wind farm (GWh)	70.44
Energy output of single CECO (GWh)	1.35
Energy output of wave farm (GWh)	13.52
Energy output of co-located wind–wave farm (GWh)	83.96

3.1.2. Energy Output: Monthly Scenario

Following the methodology presented in the previous section, the monthly energy production was computed for the stand-alone and co-located wind and wave farms. For the stand-alone wind farm, the SIMAR wind records were used to obtain the monthly Weibull probability density functions, which are shown in Figure 8. In general, the wind distribution presents a homogeneous behaviour across the different months of the year, with the most recurrent wind speeds located between 5.5 and 7 ms^{-1}. As a result, the percentage of time for which the turbine is operating, ranges from 82% in January to 70% in September. Conversely, the fraction of time for which the turbine operates at its rated power varies significantly during the year, from 12% in January and December to only 3% in September. According to Equations (3) and (7), the monthly energy production of the wind farm was computed and is presented in Figure 9. As expected, with the exception of September, a uniform monthly energy production, ranging from 7.20 GWh to 5.5 GWh, was obtained.

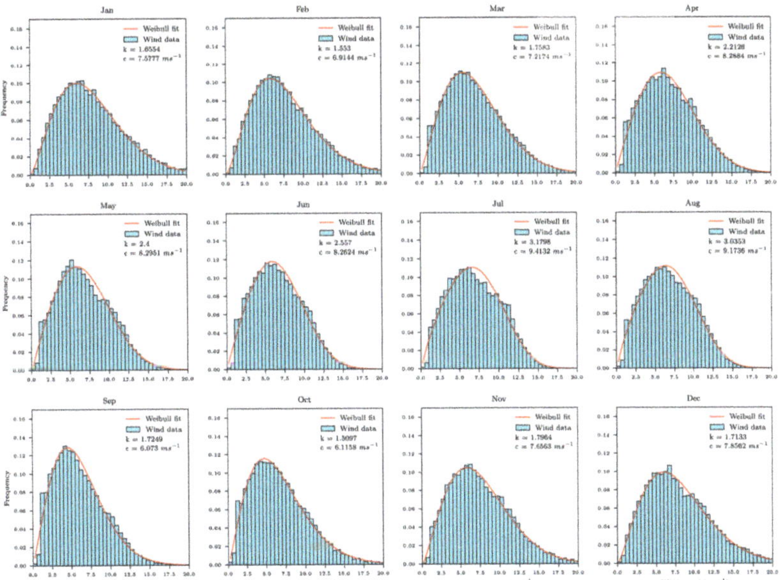

Figure 8. Observed wind data (histogram) and Weibull probability density function (red line) for the monthly scenario (k and c are the monthly scale and shape parameter values of the Weibull distribution).

For the case of the wave farm, the SIMAR wave records were used to characterise the wave resource on a monthly basis. Therefore, the omni-directional monthly wave energy matrices were constructed and plotted in Figure 10. Overall, the wave energy resource presents a strong seasonal variability. For the winter months (January, February, March and

December), the bulk of wave energy is concentrated in the range of 2–5 m of H_s and 9–15 s of T_P, with some sea states exceeding monthly energy values of 4 MWhm^{-1}. Conversely, for the summer months (from June to August), the wave energy resource is significantly lower, with a very limited number of sea states exceeding 1 MWhm^{-1}, which is mainly concentrated in the range of 1–3 m of H_s and 5–13 s of T_P. Again, combining the CECO and wave resource matrices (Equations (8) and (9)), the monthly energy output of the wave farm was computed and plotted in Figure 9. As expected, energy production presents a strong intra-annual variability, ranging from 1.57 GWh in January to 0.67 GWh in July.

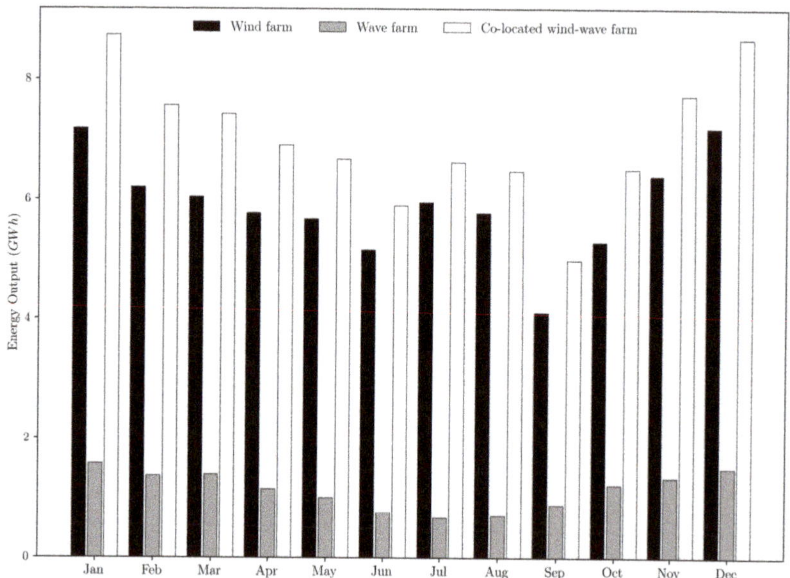

Figure 9. Monthly energy output for the stand-alone and co-located wind and wave farms.

On these grounds, the monthly energy production of the co-located wind–wave farm is presented in Figure 9. In general, the energy output varies significantly across the year. During the winter period, the production is in the range from 7.5 to 8.5 GWh, while, for the summer months is in the order of 5–6.5 GWh. This behaviour is strictly related to the strong intra-annual variability of the wave energy resource (Figure 10). For instance, during the winter months, the energy output of the wave farm represents, approximately, from 20 to 25% of the wind farm. This fact highlights the potential of CECO in terms of energy production, considering that the installed capacity of the wave farm is a fifth of the wind farm. However, for the summer months, the wave farm barely represents 10% of the wind farm energy output. All in all, the presence of the co-located wave farm would increase the overall energy production up to 23% and 10% during the winter and summer months, respectively.

In addition, the monthly variation in the capacity factor for the stand-alone and co-located wind and wave farms is shown in Figure 11. Overall, the wind farm presents a low intra-annual variability in the capacity factor, with its values ranging between 0.41 and 0.30, with the exception of September, with the lowest value of 0.24. This fact is in good agreement with the homogeneous monthly distribution of the wind resource shown in Figure 8. On the other hand, the capacity factor of the wave farm presents a strong intra-annual variability (ranging from 0.42 in the winter months to 0.17 in the summer months), which is coherent with the seasonal behaviour of the wave resource in the area of study (Figure 10). It is worth pointing out that, for the winter months, the capacity factor of the wave farm exceeds the values obtained for the wind farm. This fact is due to the relatively small electric generator used by the CECO device (500 kW). Finally, for the co-located wind–wave farm, the capacity factor is similar to the stand-alone wind farm, with the highest discrepancies present in the summer months, for which the low values of the wave farm hinder the overall capacity factor of the co-located farm.

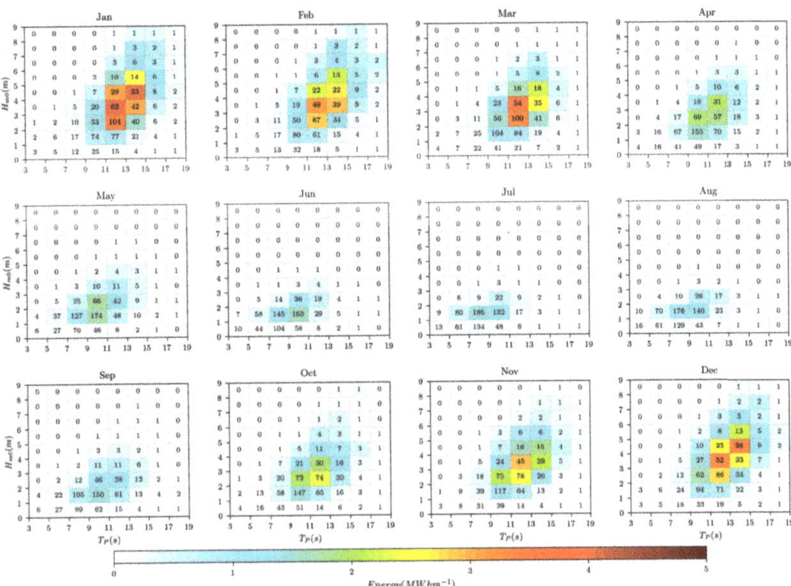

Figure 10. Wave resource energy matrix. (Colourmap represents the monthly energy per meter of wave front and the numbers represent the occurrence in annual hours for each sea state (O_i)).

3.2. Power Smoothing

From the results presented in Figures 9 and 11, it becomes apparent that the presence of the co-located wave farm would be counterproductive to reduce the variability in the intra-annual energy produced by the wind farm. In this context, the power output for the stand-alone wind farm presents a variation coefficient of $CV_{wnd} = 0.76$, while the combined wind–wave farm presents a value of $CV_{wnd-wav} = 0.75$. Therefore, according to Equation (10), the power smoothing index (PSI) would be close to zero, confirming that the co-located wave farm has no positive effects in smoothing the power output of the wind farm. This fact is mainly due to the combination of the strong seasonal behaviour of the wave resource in the area of study (Figure 10) and the present stage of development of CECO, which must be optimised to improve its efficiency and energy production for milder wave conditions [46].

3.3. LCoE Analysis

The LCoE associated with the wind, wave and co-located farms was assessed using the analytical cost model proposed by Clark et al. [30], which is described in detail in Section 2.3.3. In this context, Table 8 summarises the main input values of the LCoE model used in the present work. It is important to note that the most conservative values proposed by Clark et al. [30] were taken for parameters, such as inflation, loan, insurance and decommissioning rates.

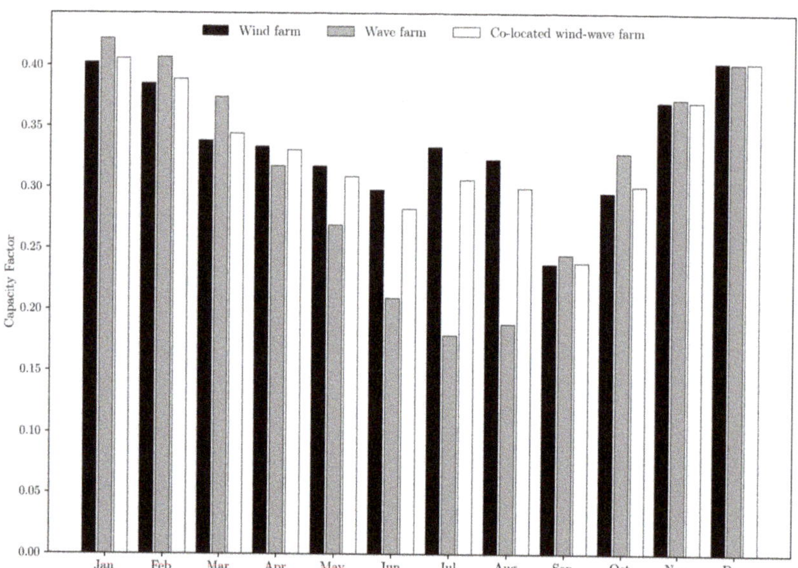

Figure 11. Monthly capacity factor for the stand-alone and co-located wind and wave farms.

Table 8. Input parameters for the LCoE model.

Parameter	Value
Lifespan (years)	20
Inflation rate, $r_{inflation}$ (%)	2
Loan rate, r_{loan} (%)	10
Insurance rate, r_{ins} (%)	2
Decommissioning rate, r_{decm} (%)	3
Length of mooring wind farm (m)	2700
Length of mooring wave farm (m)	6252
Length of mooring co-located farm (m)	8952
Length of inter-array cable wind farm (km)	1.2
Length of inter-array cable wave farm (km)	2.8
Length of inter-array cable co-located farm (km)	5.2
Length of export cable (km)	20

For the stand-alone wind farm, an LCoE of 0.096 USD/kWh was obtained, which is in good agreement with the values reported by the floating offshore wind energy industry [66,67]. Figure 12 summarises the pre-installation, implementation, operation and decommissioning costs per megawatt of installed capacity. As expected, the implementation and operation stages account for the majority of the total costs, representing 41% and 52%, respectively. The remainder costs are split between the pre-installation (4.47%) and decommissioning (2.92%) stages.

On the other hand, the LCoE obtained for the stand-alone wave farm was 0.347 USD/kWh, which is in line with previous works dealing with the LCoE assessment of wave energy farms [30]. In comparison with the wind farm, the total cost, per megawatt of installed capacity, would increase by 68%. This fact is explained by the low maturity level of the wave energy industry, which is especially relevant for the installation, operation and maintenance tasks (Figure 12). In consequence, it can be concluded that a CECO wave farm is not yet ready for full commercial exploitation. Nonetheless, CECO is still at an early stage of development; therefore, significant LCoE reductions could be achieved by the optimisation of its design and the economy-of-scale effect.

Finally, the co-located wave–wind farm yields an LCoE of 0.115 USD/kWh, which represents a substantial reduction (up to 200%) in comparison with the stand-alone wave farm. As can be observed in Figure 12, the largest savings are concentrated in the implementation and operation costs. On the one hand, for the implementation stage, expenses are reduced by 39%, taking advantage of the shared electrical infrastructures such as the offshore and onshore substations and the export cable. On the other hand, operation costs present reductions in the order of 44%, as a result of shared administration, operation, maintenance, transport, insurance and facility costs. Considering the total cost per megawatt of installed capacity, the results are even more encouraging. In this context, the co-located wave–wind farm would present a similar total cost (per megawatt) to that of the stand-alone wind farm and a reduction of 41% in comparison with the stand-alone wave farm. Consequently, from the economic point of view, CECO appears as a feasible option to be used in a co-located wave–wind farm.

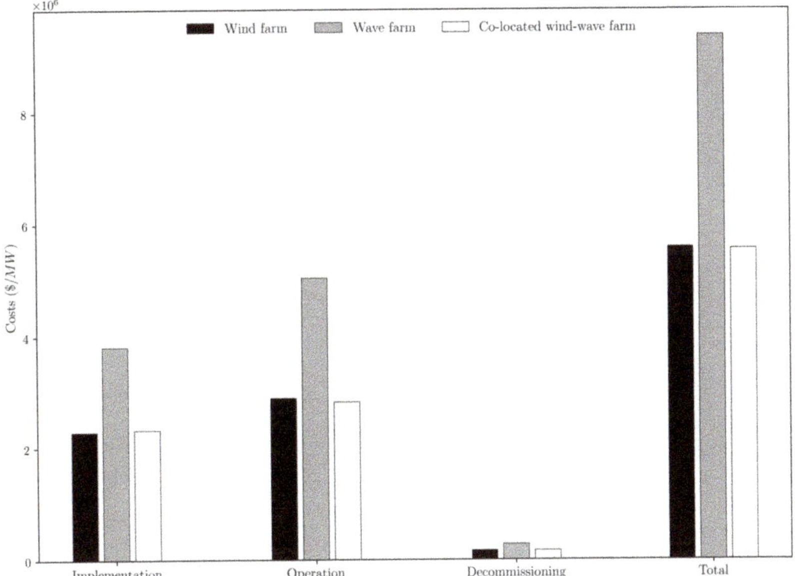

Figure 12. Costs per megawatt of the stand-alone and co-located wind and wave farms.

4. Discussion

This section presents a discussion related to the main aspects tackled in this investigation, namely, (i) the comparative analysis between the stand-alone wave and wind farms and (ii) the potential benefits derived from the co-located wave–wind farm.

First of all, the results obtained in Sections 3.1.1 and 3.1.2 confirm the promising potential of the CECO concept to harvest wave energy. In annual terms, the energy output, per megawatt of installed capacity, of CECO is larger than the one produced by the offshore wind turbine (Table 7). This fact is especially noticeable during the winter months (Figure 9). It is noteworthy to point out that the annual performance of CECO is clearly hindered by the strong seasonal variability in the wave resource in the area of study (Figure 10). Therefore, in locations with a more homogeneous intra-annual wave resource, the performance of CECO could be even better. In terms of LCoE, the values obtained for the stand-alone CECO farm are aligned with the trends observed for the wave energy industry (i.e., approximately ten times higher than traditional renewable sources). Consequently, at its current stage,

CECO is not viable from the commercial point of view. Nonetheless, CECO is still at an early stage of development; therefore, the design optimisation of mooring systems, device geometry, PTO configuration and control strategies may result in significant reductions in the associated LCoE. Furthermore, mass production of the CECO device would contribute to decreasing, even more, its associated costs due to the economy-of-scale effect.

On the other hand, the results obtained proved that, in combination with offshore wind farms, CECO could progress more rapidly towards commercialisation. In comparison with the stand-alone CECO farm, the LCoE value obtained for the co-located wave–wind farm (0.115 USD/kWh) is almost three times lower. This value is similar to the LCoE of the offshore wind industry as of five years ago [8]. Therefore, the proposed co-located CECO–wind farm would be on the edge of commercial viability. These results appear even more encouraging when considering that a preliminary layout was used for the CECO farm, since its optimum design is beyond the scope of this work. Furthermore, it is worth noting that the LCoE model [30] was applied without considering any discount method, such as the so-called learning rates, and taking conservative values for parameters, such as inflation, loan, insurance and decommissioning rates. On the other hand, it is important to point out that the cost model developed by Clark et al. (2019) presents certain limitations, which may contribute to uncertainties in the LCoE estimations [30]. Therefore, the results presented in the present work should be taken as a first approximation and a more accurate estimation of the LCoE would require the use of a higher TRL version of CECO and a technology-specific LCoE model. Besides the LCoE reductions observed, the co-location of the wave array would also increase the annual energy output by 19%. However, the presence of the co-located wave array would not contribute to reducing the intra-annal variability of the energy produced by the wind farm. As aforementioned, the power smoothing capacity of the wave farm is hindered by the strong seasonal variations of the wave resource in the area of study (Figure 10). Nonetheless, it is important to point out that the contribution of CECO to power smoothing could be significantly improved by device optimisation and the adoption of control strategies, which may lead to an increase in the energy production for milder wave conditions, consequently reducing the intra-annual power variability. Another aspect that may have influence on the results obtained in terms of power smoothing is the design considered for the wave farm. As indicated in Section 2.2.2, several constraints were considered for its definition, including the use of a single row of devices, with a conservative separation of 330 m, restricted to the sea space allocated for the offshore wind farm. As a result, the layout obtained consisted of 10 CECO units (Figure 5), with an installed capacity of 5 MW, and should be considered as a preliminary design. In this context, the use of an optimised layout with a larger installed capacity could increase the energy output generated for milder wave conditions, consequently resulting in a greater contribution to smooth the power output of the wind farm. However, this fact should be corroborated with future research. Finally, additional benefits could derive from the presence of the co-located wave farm, such as milder wave conditions within the area of the wind farm, which could facilitate accessibility, operation and maintenance tasks.

5. Conclusions

In recent years, the interest in exploiting the energy resource of ocean waves has translated into the development of multiple WEC concepts. However, WEC technology is still in its infancy, with associated LCoE values up to ten times higher than traditional renewable sources, therefore, far from commercial viability. In this context, potential synergies with the offshore wind energy industry could significantly contribute towards the development of a fully fledged wave energy industry. Among them, the share of operational and capital costs and of power output smoothing stands out. On the other hand, offshore wind farms could also benefit from the sheltering effect of co-located wave farms, mitigating harsh wave conditions within the farm, consequently facilitating operation and maintenance tasks. On these grounds, the objective of this work is to assess the potential benefits of co-locating the CECO device, a promising WEC concept that uses a sloped PTO configuration to harness both the kinetic and potential energy of the waves, with the commercial *WindFloat Atlantic* offshore wind farm.

Overall, CECO presents promising results in terms of energy production (per megawatt of installed capacity), outperforming the offshore wind turbine, especially during the winter months. However, CECO is still far from commercial feasibility, presenting an associated LCoE of 0.347 USD/kWh, which is approximately 4 and 7 times higher than the values of the offshore and onshore wind industries, respectively. In consequence, significant design refinement dealing with mooring systems, device geometry, PTO configuration and control strategies must be conducted to reduce the associated LCoE of CECO. Furthermore, the results show that the co-location with offshore wind farms could facilitate the transition of CECO towards commercialisation. In this sense, the LCoE obtained for the co-located CECO-wind farm was 0.115 USD/kWh, which is almost three times lower than the value of stand-alone CECO wave farms and similar to the levels of the offshore wind industry as of five years ago. Finally, the expected know-how acquired from the offshore energy industry could also help to improve the efficiency of transport, deployment, operational and maintenance tasks of CECO farms.

In summary, the results presented in this paper highlight the benefits of co-locating the CECO device with an offshore wind farm. In addition, this study presents a benchmark to compare, in the same sea space, the performance of a WEC technology with a commercial offshore wind turbine. Lastly, it is worth mentioning that additional aspects of the CECO farm, including its optimum layout, associated shadow effects and impact on temporal windows for operation and maintenance tasks, are beyond the scope of this paper; therefore, they are to be addressed in detail in future research.

Author Contributions: Conceptualization, V.R., G.G., T.C.-C., M.L., P.R.-S. and F.T.-P.; methodology, V.R., G.G., T.C.-C., M.L., P.R.-S. and F.T.-P.; software, V.R., G.G. and M.L.; resources, V.R.; writing—original draft preparation, V.R.; writing—review and editing, V.R., G.G., T.C.-C., M.L., P.R.-S. and F.T.-P.; funding acquisition, V.R., M.L., P.R.-S. and F.T.-P. All authors have read and agreed to the published version of the manuscript.

Funding: This research study was partially funded by the *Ports Towards Energy Self-Sufficiency* (PORTOS) project co-financed by the Interreg Atlantic Area Programme through the European Regional Development Fund, grant number EAPA-784/2018. Furthermore, during this research study, Victor Ramos was supported by the program of *Stimulus of Scientific Employment Individual Support* (CEECIND/03665/2018) from the Portuguese Foundation of Science and Technology (FCT).

Institutional Review Board Statement: Not applicable.

Informed Consent Statement: Not applicable.

Data Availability Statement: Not applicable.

Acknowledgments: The authors would like to thank the Spanish Port Authority (Puertos del Estado) for providing the wind and wave datasets required for the elaboration of the present work.

Conflicts of Interest: The authors declare no conflicts of interest.

Abbreviations

The following abbreviations are used in this manuscript:

EU	European Union
MRE	Marine renewable energy
WEC	Wave energy converter
LCoE	Levelised cost of energy
NAO	North atlantic oscillation
PTO	Power take-off
LLM	Lateral mobile module
TRL	Technology readiness level

References

1. United Nations. *Adoption of the Paris Agreement. United Nations Framework Convention on Climate Change (UNFCCC)*; United Nations: San Francisco, CA, USA, 2015.
2. United Nations. *Transforming Our World, the 2030 Agenda for Sustainable Development*; United Nations: San Francisco, CA, USA, 2015.
3. Maris, G.; Flouros, F. The Green Deal, National Energy and Climate Plans in Europe: Member States Compliance and Strategies. *Adm. Sci.* **2021**, *11*, 75. [CrossRef]
4. Melikoglu, M. Current status and future of ocean energy sources: A global review. *Ocean. Eng.* **2018**, *148*, 563–573. [CrossRef]
5. Ocean Energy Forum. European Commission. Ocean Energy Strategic Roadmap. Building Ocean Eenergy for Europe. 2016. Available online: https://webgate.ec.europa.eu/maritimeforum/en/node/3962 (accessed on 22 January 2018).
6. Perez-Collazo, C.; Greaves, D.; Iglesias, G. A review of combined wave and offshore wind energy. *Renew. Sustain. Energy Rev.* **2015**, *42*, 141–153. [CrossRef]
7. Global Wind Energy Council, GWEC. Global Wind Report 2021. Available online: https://gwec.net/wp-content/uploads/2021/03/GWEC-Global-Wind-Report-2021.pdf (accessed on 10 November 2021).
8. International Renewable Energy Agency (IRENA). Renewable Capacity Statistics 2021. Available online: https://www.irena.org/publications/2021/March/Renewable-Capacity-Statistics-2021 (accessed on 10 November 2021).
9. De Castro, M.; Salvador, S.; Gómez-Gesteira, M.; Costoya, X.; Carvalho, D.; Sanz-Larruga, F.; Gimeno, L. Europe, China and the United States: Three different approaches to the development of offshore wind energy. *Renew. Sustain. Energy Rev.* **2019**, *109*, 55–70. [CrossRef]
10. International Renewable Energy Agency (IRENA). Renewable Power Generation Costs in 2020. Available online: https://www.irena.org/publications/2021/Jun/Renewable-Power-Costs-in-2020 (accessed on 10 November 2021).
11. International Renewable Energy Agency (IRENA). Floating Foundations: A Game Changer for Offshore Wind. Available online: https://www.irena.org/publications/2016/Dec/Floating-foundations-A-game-changer-for-offshore-wind (accessed on 10 November 2021).
12. Martinez, A.; Iglesias, G. Wave exploitability index and wave resource classification. *Renew. Sustain. Energy Rev.* **2020**, *134*, 110393. [CrossRef]
13. López, I.; Andreu, J.; Ceballos, S.; de Alegría, I.M.; Kortabarria, I. Review of wave energy technologies and the necessary power-equipment. *Renew. Sustain. Energy Rev.* **2013**, *27*, 413–434. [CrossRef]
14. Todalshaug, J.H.; Ásgeirsson, G.S.; Hjálmarsson, E.; Maillet, J.; Möller, P.; Pires, P.; Guérinel, M.; Lopes, M. Tank testing of an inherently phase-controlled wave energy converter. *Int. J. Mar. Energy* **2016**, *15*, 68–84. [CrossRef]
15. Giannini, G.; Day, S.; Rosa-Santos, P.; Taveira-Pinto, F. A Novel 2-D Point Absorber Numerical Modelling Method. *Inventions* **2021**, *6*, 75. [CrossRef]
16. Kofoed, J.P.; Frigaard, P.; Friis-Madsen, E.; Sørensen, H.C. Prototype testing of the wave energy converter wave dragon. *Renew. Energy* **2006**, *31*, 181–189. [CrossRef]
17. Fernandez, H.; Iglesias, G.; Carballo, R.; Castro, A.; Fraguela, J.; Taveira-Pinto, F.; Sanchez, M. The new wave energy converter WaveCat: Concept and laboratory tests. *Mar. Struct.* **2012**, *29*, 58–70. [CrossRef]
18. Heath, T. A review of oscillating water columns. *Philos. Trans. R. Soc. Math. Phys. Eng. Sci.* **2012**, *370*, 235–245. [CrossRef] [PubMed]
19. Calheiros-Cabral, T.; Clemente, D.; Rosa-Santos, P.; Taveira-Pinto, F.; Ramos, V.; Morais, T.; Cestaro, H. Evaluation of the annual electricity production of a hybrid breakwater-integrated wave energy converter. *Energy* **2020**, *213*, 118845. [CrossRef]
20. Henderson, R. Design, simulation, and testing of a novel hydraulic power take-off system for the Pelamis wave energy converter. *Renew. Energy* **2006**, *31*, 271–283. [CrossRef]
21. Wave energy device Oyster launched. *Renew. Energy Focus* **2009**, *10*, 14. [CrossRef]
22. Carnegie Clean Energy. CETO Technology. Available online: https://www.carnegiece.com/ceto-technology/ (accessed on 22 January 2020).
23. Aderinto, T.; Li, H. Ocean Wave Energy Converters: Status and Challenges. *Energies* **2018**, *11*, 1250. [CrossRef]
24. European Commission. Working Group Ocean Energy. SET-Plan Ocean Energy Implementation Plan. 2018. Available online: https://setis.ec.europa.eu/system/files/set_plan_ocean_implementation_plan.pdf (accessed on 12 September 2020).
25. Choupin, O.; Pinheiro Andutta, F.; Etemad-Shahidi, A.; Tomlinson, R. A decision-making process for wave energy converter and location pairing. *Renew. Sustain. Energy Rev.* **2021**, *147*, 111225. [CrossRef]
26. Astariz, S.; Iglesias, G. Enhancing Wave Energy Competitiveness through Co-Located Wind and Wave Energy Farms. A Review on the Shadow Effect. *Energies* **2015**, *8*, 7344–7366. [CrossRef]
27. Astariz, S.; Iglesias, G. The economics of wave energy: A review. *Renew. Sustain. Energy Rev.* **2015**, *45*, 397–408. [CrossRef]
28. Astariz, S.; Perez-Collazo, C.; Abanades, J.; Iglesias, G. Co-located wind-wave farm synergies (Operation & Maintenance): A case study. *Energy Convers. Manag.* **2015**, *91*, 63–75. [CrossRef]
29. Astariz, S.; Perez-Collazo, C.; Abanades, J.; Iglesias, G. Co-located wave-wind farms: Economic assessment as a function of layout. *Renew. Energy* **2015**, *83*, 837–849. [CrossRef]
30. Clark, C.E.; Miller, A.; DuPont, B. An analytical cost model for co-located floating wind-wave energy arrays. *Renew. Energy* **2019**, *132*, 885–897. [CrossRef]

31. Astariz, S.; Abanades, J.; Perez-Collazo, C.; Iglesias, G. Improving wind farm accessibility for operation & maintenance through a co-located wave farm: Influence of layout and wave climate. *Energy Convers. Manag.* **2015**, *95*, 229–241. [CrossRef]

32. Astariz, S.; Iglesias, G. Co-located wind and wave energy farms: Uniformly distributed arrays. *Energy* **2016**, *113*, 497–508. [CrossRef]

33. Astariz, S.; Iglesias, G. Selecting optimum locations for co-located wave and wind energy farms. Part II: A case study. *Energy Convers. Manag.* **2016**, *122*, 599–608. [CrossRef]

34. Stoutenburg, E.D.; Jenkins, N.; Jacobson, M.Z. Power output variations of co-located offshore wind turbines and wave energy converters in California. *Renew. Energy* **2010**, *35*, 2781–2791. [CrossRef]

35. Fusco, F.; Nolan, G.; Ringwood, J.V. Variability reduction through optimal combination of wind/wave resources—An Irish case study. *Energy* **2010**, *35*, 314–325. [CrossRef]

36. Cradden, L.; Mouslim, H.; Duperray, O.; Ingram, D. Joint exploitation of wave and offshore wind power. In Proceedings of the Nineth European Wave and Tidal Energy Conference (EWTEC), Southampton, UK, 7 September 2011; pp. 5–9.

37. Astariz, S.; Iglesias, G. Output power smoothing and reduced downtime period by combined wind and wave energy farms. *Energy* **2016**, *97*, 69–81. [CrossRef]

38. Gaughan, E.; Fitzgerald, B. An assessment of the potential for Co-located offshore wind and wave farms in Ireland. *Energy* **2020**, *200*, 117526. [CrossRef]

39. Astariz, S.; Iglesias, G. Selecting optimum locations for co-located wave and wind energy farms. Part I: The Co-Location Feasibility index. *Energy Convers. Manag.* **2016**, *122*, 589–598. [CrossRef]

40. Astariz, S.; Vazquez, A.; Sánchez, M.; Carballo, R.; Iglesias, G. Co-located wave-wind farms for improved O&M efficiency. *Ocean. Coast. Manag.* **2018**, *163*, 66–71. [CrossRef]

41. Astariz, S.; Iglesias, G. Accessibility for operation and maintenance tasks in co-located wind and wave energy farms with non-uniformly distributed arrays. *Energy Convers. Manag.* **2015**, *106*, 1219–1229. [CrossRef]

42. Ramos, V.; López, M.; Taveira-Pinto, F.; Rosa-Santos, P. Performance assessment of the CECO wave energy converter: Water depth influence. *Renew. Energy* **2018**, *117*, 341–356. [CrossRef]

43. Rosa-Santos, P.; Taveira-Pinto, F.; Rodríguez, C.A.; Ramos, V.; López, M. The CECO wave energy converter: Recent developments. *Renew. Energy* **2019**, *139*, 368–384. [CrossRef]

44. EDP Renewables. WindFloat Atlantic Project. 2019. Available online: https://www.edpr.com/en/news/2019/10/21/windfloat-atlantic-begins-installation-first-floating-wind-farm (accessed on 20 January 2020).

45. Salvação, N.; Soares, C.G. Wind resource assessment offshore the Atlantic Iberian coast with the WRF model. *Energy* **2018**, *145*, 276–287. [CrossRef]

46. Ramos, V.; López, M.; Taveira-Pinto, F.; Rosa-Santos, P. Influence of the wave climate seasonality on the performance of a wave energy converter: A case study. *Energy* **2017**, *135*, 303–316. [CrossRef]

47. EDP Renewables. WindFloat Atlantic Project. 2019. Available online: https://www.edpr.com/en/news/2019/12/30/second-platform-windfloatatlantic-has-set-port-ferrol (accessed on 11 October 2021).

48. Silva, D.; Martinho, P.; Guedes Soares, C. Wave energy distribution along the Portuguese continental coast based on a thirty three years hindcast. *Renew. Energy* **2018**, *127*, 1064–1075. [CrossRef]

49. Rosa-Santos, P.; Taveira-Pinto, F.; Teixeira, L.; Ribeiro, J. CECO wave energy converter: Experimental proof of concept. *J. Renew. Sustain. Energy* **2015**, *7*, 061704. [CrossRef]

50. Rosa-Santos, P.; Taveira-Pinto, F.; Pinho-Ribeiro, J.; Teixeira, L.; Marinheiro, J. Harnessing the kinetic and potential wave energy: Design and development of a new wave energy converter. In *Renewable Energies Offshore*; CRC Press: Boca Raton, FL, USA, 2015; pp. 367–374.

51. López, M.; Taveira-Pinto, F.; Rosa-Santos, P. Numerical modelling of the CECO wave energy converter. *Renew. Energy* **2017**, *113*, 202–210. [CrossRef]

52. Rodríguez, C.A.; Rosa-Santos, P.; Taveira-Pinto, F. Hydrodynamic optimization of the geometry of a sloped-motion wave energy converter. *Ocean. Eng.* **2020**, *199*, 107046. [CrossRef]

53. Giannini, G.; López, M.; Ramos, V.; Rodríguez, C.A.; Rosa-Santos, P.; Taveira-Pinto, F. Geometry assessment of a sloped type wave energy converter. *Renew. Energy* **2021**, *171*, 672–686. [CrossRef]

54. Rodríguez, C.A.; Rosa-Santos, P.; Taveira-Pinto, F. Assessment of damping coefficients of power take-off systems of wave energy converters: A hybrid approach. *Energy* **2019**, *169*, 1022–1038. [CrossRef]

55. López, M.; Ramos, V.; Rosa-Santos, P.; Taveira-Pinto, F. Effects of the PTO inclination on the performance of the CECO wave energy converter. *Mar. Struct.* **2018**, *61*, 452–466. [CrossRef]

56. Giannini, G.; Rosa-Santos, P.; Ramos, V.; Taveira-Pinto, F. On the Development of an Offshore Version of the CECO Wave Energy Converter. *Energies* **2020**, *13*, 1036. [CrossRef]

57. REVE News of the Wind Sector in Spain and in the World. Floating Wind Energy: First Wind Turbine of WindFloat Atlantic Moves Into Position. 2019. Available online: https://www.evwind.es/2019/10/22/floating-wind-energy-first-wind-turbine-of-windfloat-atlantic-moves-into-position/71445 (accessed on 11 October 2021).

58. Principle Power Inc. The WindFloat Advantage. 2019. Available online: https://www.principlepower.com/windfloat/the-windfloat-advantage (accessed on 11 October 2021).

59. EDP Renewables. WindFloat Atlantic Project. 2021. Available online: https://www.edpr.com/en/innovation (accessed on 11 October 2021).
60. Vestas V164-8.0 MW Technical Specifcations. 2021. Available online: https://pdf.archiexpo.com/pdf/vestas/vestas-v164-80-mw/88087-134417.html (accessed on 11 October 2021).
61. Astariz, S.; Perez-Collazo, C.; Abanades, J.; Iglesias, G. Towards the optimal design of a co-located wind-wave farm. *Energy* **2015**, *84*, 15–24. [CrossRef]
62. López, M.; Rodríguez, N.; Iglesias, G. Combined Floating Offshore Wind and Solar PV. *J. Mar. Sci. Eng.* **2020**, *8*, 576. [CrossRef]
63. Puertos del Estado. SIMAR Dataset. 2020. Available online: https://bancodatos.puertos.es//BD/informes/INT_8.pdf (accessed on 11 October 2021).
64. Myhr, A.; Bjerkseter, C.; Ågotnes, A.; Nygaard, T.A. Levelised cost of energy for offshore floating wind turbines in a lifeăcycle perspective. *Renew. Energy* **2014**, *66*, 714–728. [CrossRef]
65. EDP Renewables. WindFloat Atlantic Project. 2021. Available online: https://www.edpr.com/en/news/2021/09/23/windfloat-atlantic-reaches-75-gwh-its-first-year-operation (accessed on 11 October 2021).
66. Kausche, M.; Adam, F.; Dahlhaus, F.; Großmann, J. Floating offshore wind—Economic and ecological challenges of a TLP solution. *Renew. Energy* **2018**, *126*, 270–280. [CrossRef]
67. Lerch, M.; De-Prada-Gil, M.; Molins, C.; Benveniste, G. Sensitivity analysis on the levelized cost of energy for floating offshore wind farms. *Sustain. Energy Technol. Assessments* **2018**, *30*, 77–90. [CrossRef]

Journal of
Marine Science and Engineering

Article

Flume Experiments on Energy Conversion Behavior for Oscillating Buoy Devices Interacting with Different Wave Types

Shufang Qin [1,2], Jun Fan [1,2,*], Haiming Zhang [1,2], Junwei Su [1,2] and Yi Wang [1,2]

1   Key Laboratory of Ministry of Education for Coastal Disaster and Protection, Hohai University, Nanjing 210024, China; sfqin@hhu.edu.cn (S.Q.); haiming@hhu.edu.cn (H.Z.); 171303020052@hhu.edu.cn (J.S.); shgovwy@hotmail.com (Y.W.)
2   College of Harbour, Coastal and Offshore Engineering, Hohai University, Nanjing 210024, China
*   Correspondence: fanjun@hhu.edu.cn; Tel.: +86-137-7668-1019

Abstract: Oscillating buoy device, also known as point absorber, is an important wave energy converter (WEC) for wave energy development and utilization. The previous work primarily focused on the optimization of mechanical design, buoy's array configuration and the site selection with larger wave energy density in order to improve the wave energy generation performance. In this work, enlightened by the potential availability of Bragg reflection induced by multiple submerged breakwaters in nearshore areas, we investigate the energy conversion behavior of oscillating buoy devices under different wave types (traveling waves, partial and fully standing waves) by flume experiments. The localized partial standing wave field is generated by the Bragg resonance at the incident side of rippled bottoms. Furthermore, the fully standing wave field is generated by the wave reflection of vertical baffle installed in flume. Then the wave power generation performance is discussed under the conditions with the same wave height but different wave types. The experimental measurements show that the energy conversion performance of the oscillating buoy WEC could be improved under the condition of standing waves when compared with traveling waves. This work provides the experimental comparison evidence of wave energy conversion response of oscillating buoy devices between travelling waves and standing (fully or partial) wave conditions.

Keywords: wave power; oscillating buoy; power generation performance; standing waves; experimental research

Citation: Qin, S.; Fan, J.; Zhang, H.; Su, J.; Wang, Y. Flume Experiments on Energy Conversion Behavior for Oscillating Buoy Devices Interacting with Different Wave Types. *J. Mar. Sci. Eng.* **2021**, *9*, 852. https://doi.org/10.3390/jmse9080852

Academic Editors: Paulo Jorge Rosa-Santos, Francisco Taveira Pinto, Mario López Gallego and Claudio Alexis Rodríguez Castillo

Received: 10 July 2021
Accepted: 5 August 2021
Published: 8 August 2021

Publisher's Note: MDPI stays neutral with regard to jurisdictional claims in published maps and institutional affiliations.

1. Introduction

Wave energy is a kind of marine renewable energy, which has attracted much attention because of its merits of high energy density, being clean and renewable and having great potential for development. It has been explored for nearly 200 years [1–4]. For wave energy utilization, the basic principle is to use the wave energy converters (WECs) of various mechanisms to extract the mechanical energy of the wave including the potential energy and the kinetic energy to generate electricity [5–7]. At present, varieties of wave energy conversion devices have been developed all over the world, whose principle and structure are simple and convenient for large-scale production, and the energy conversion efficiency of wave energy generation devices is higher than that of other renewable energy sources. Therefore, the use of wave power generation as a means of energy supply in coastal areas has a wide application prospect [8–10].

On the basis of the absorption type, WEC can be primarily classified into oscillating water columns (OWCs), overtopping or terminator devices, attenuators, oscillating buoy devices (point absorbers), oscillating wave surge, submerged pressure differential and rotating mass [11–18]. Oscillating buoy device is a hot spot in recent years, and it is also the most probable wave device for commercialization. The device absorbs wave energy by means of a cylinder or sphere placed in water as a carrier for energy absorption, depending on the relative motion between the floating body and the device, and then drives the mechanical system or hydraulic system connected with the buoy to run the generator to generate electrical energy [19–22]. The key feature of oscillating buoy device is the high energy conversion rate and stable power generation process. The buoy has low production cost and simple construction process [23]. The oscillating buoy WEC has entered the stage of commercial power generation in developed countries in Europe and America. It was originally a fixed Archimedes Wave Swing (AWS) device developed by Amsterdam Company of the Netherlands. The device was later modified by the British AWS Ocean Energy company into a floating Archimedes power plant. The Power Buoy device developed by OPT (Ocean Power Technologies) in the United States has been successfully tested in the Pacific Ocean and Atlantic Ocean. British Ocean Navitas Limited Company has successfully developed Aegir Dynamo wave generator unit set with a pontoon structure [24,25]. In China, Shi [26], Cheng [27] and Gou [28] et al. have independently studied the oscillating buoy devices and conducted sea tests. The performance of a single buoy is always limited. As the scale of power generation increases, it is a trend in this field to combine the buoys and arrange them in an array to coordinate power generation. Therefore, the concept of array optimization is proposed, which uses a group of buoys to form an array to complete the unified extraction and transformation of wave energy. Nielsen [29], Tokić [30], Folley [31], Do [32], Amini [33] and Neshat [34] have carried out in-depth studies on the layout optimization of oscillating buoys.

According to the research status and development trend of the world, the current research on wave power generation is mainly focused on how to effectively promote the specific mechanical design of wave energy conversion devices in the fields of energy transfer and conversion performance and the combined wave energy generation device. The layout optimization, microgrid and other fields have also been studied in depth. To the best of our knowledge, however, there are few research results to improve the performance of wave energy generation by increasing the wave energy density of the target sea area. What is more, in the previous experimental research on wave power generation with oscillating buoys, the traveling waves are the main concerns in wave type, while there is no research on the characteristics of power generation equipment under the condition of fully standing waves or partial standing waves.

Tao [35] preliminarily studied the behavior of wave energy generation enhancement by Bragg resonance through physical model experiments, that is, through the interaction between incident waves and periodic wavy bottom. Thus, the wave energy density of the target region is increased. The results show that, compared with the wave power generated by the oscillating buoy (point absorption) device on flat bottom, the Bragg resonance energy can cause the power to increase by more than 10 times. By analyzing the reason for the improvement, it is also found that the increase of generation power is not only due to the increase of wave energy density but also to the wave types. This is because when the Bragg resonance occurs, the superposition of incident waves and reflected waves forms standing waves, which causes the motion of water particles under the wave action to change from the approximately closed elliptical trajectory to an approximate straight line, and water particles at the wave antinodes only move up and down vertically. Thus, the stability of the oscillating buoy is enhanced, and the mechanical resonance of oscillating buoy along with water particles could be enhanced, which will affect the generation performance of the WEC.

For further investigation of the effects of wave types on wave energy conversion by oscillating buoy on the basis of previous research above, series wave flume experiments are performed to investigate the effects of different wave types at the same wave height on the performance of wave energy generation. The main wave types considered are traveling waves, partial standing waves and fully standing waves. This study aims to deepen and enrich our understanding about the wave energy generation. It provides the experimental evidence to compare the wave energy conversion response of oscillating buoy devices between travelling waves and standing (fully or partial) wave conditions.

The remainder of this paper is organized as follows. The motion characteristics of water particles under traveling waves, partial standing waves and fully standing waves are briefly described respectively in Section 2. The experimental design and model layout are presented in Section 3. Results and discussions are provided in Section 4. Concluding remarks are summarized in Section 5.

2. Characteristics of Water Particles Motion of Different Types of Waves

In the Airy wave, the horizontal and vertical displacement of the water particle at (x_0, z_0) under travelling waves can be described as below [36]:

$$\xi = \int_0^t \frac{\partial \phi}{\partial x} dt = -A \frac{\cosh k(z_0 + h)}{\sinh kh} \sin(kx_0 - \sigma t) \tag{1}$$

$$\zeta = \int_0^t \frac{\partial \phi}{\partial z} dt = A \frac{\sinh k(z_0 + h)}{\sinh kh} \cos(kx_0 - \sigma t) \tag{2}$$

in which, A is the wave amplitude of travelling wave. k is the wave number and h is the water depth. σ is the circular frequency.

It is evident that the particle orbit of traveling waves is a closed ellipse (circle in deep water), which is a combination of both horizontal (amplitude $A \frac{\cosh k(z_0+h)}{\sinh kh}$) and vertical (amplitude $A \frac{\sinh k(z_0+h)}{\sinh kh}$) wave motions. The orbit of the water particle on free surface at any position is the same.

Fully standing waves refer to a kind of wave superimposed when two traveling waves are opposite in direction while same in wave amplitude (expressed still as A) and wave period, the horizontal and vertical displacement of water particle under fully standing waves can be written as [36]:

$$\xi = -2A \frac{\cosh k(z_0 + h)}{\sinh kh} \sin kx_0 \cos \sigma t \tag{3}$$

$$\zeta = 2A \frac{\sinh k(z_0 + h)}{\sinh kh} \cos kx_0 \cos \sigma t \tag{4}$$

It should be noted that the amplitude parameter A is still the original amplitude of travelling waves of either direction. After the superposition of two opposite travelling wave with same amplitude A, the values of horizontal and vertical displacements will be doubled compared with the travelling wave component. Furthermore, ξ and ζ are dependent on the position (x_0, z_0). The motion of water particle is oscillating in the vertical direction at the wave antinodes ($x_0 = n\pi/k$, $n = 0, 1, 2, 3$) and in the horizontal direction at the wave nodes ($x_0 = (n + 1/2)\pi/k$, $n = 0, 1, 2, 3$). The condition of generating above-mentioned fully standing waves is that traveling waves are directly incident to the vertical wall and completely reflected. The horizontal velocity u and the vertical velocity w of the water particle are [36]:

$$u = \frac{\partial \phi}{\partial x} = 2A\sigma \frac{\cosh k(z_0 + h)}{\sinh kh} \sin kx_0 \sin \sigma t \tag{5}$$

$$w = \frac{\partial \phi}{\partial z} = -2A\sigma \frac{\sinh k(z_0 + h)}{\sinh kh} \cos kx_0 \sin \sigma t \tag{6}$$

Similarly to displacement expressions, at wave antinodes, the horizontal velocity u of water particles is constant to zero, and the amplitude of vertical velocity and water surface fluctuation reaches the maximum. At wave nodes, the horizontal velocity u has the maximum amplitude, while the vertical velocity and the amplitude of the water surface fluctuation are always zero.

When traveling waves are not completely reflected, the superposition of incident and reflected waves is called partial standing waves. Assuming the amplitude of the incident waves is A_1, the amplitude of the reflected waves A_2 will be less than A_1, the combination of these two waves gives

$$\eta = (A_1 + A_2)\cos(kx)\cos(\sigma t) + (A_1 - A_2)\sin(kx)\sin(\sigma t) \tag{7}$$

The amplitude is $A^*_{max} = A_1 + A_2$ at antinodes and is $A^*_{min} = A_1 - A_2$ at nodes.

The general sketches of the travelling waves, partial standing waves and fully standing waves are presented in Figure 1 as below.

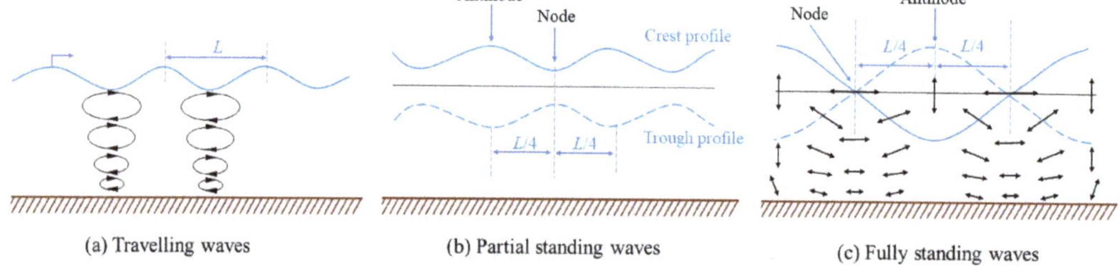

(a) Travelling waves   (b) Partial standing waves   (c) Fully standing waves

Figure 1. Sketch of the three types of waves.

It is worth noting that when the incident waves pass through the underwater sinusoidal topography and satisfy the topography wavelength as the integer multiple of half wavelength of the surface wave, it will induce the wave Bragg resonance phenomenon. The incident waves will be reflected and superimposed on the incident waves as partial standing waves.

3. Experimental Design and Model Layout

Based on the existing knowledge of water particles' motion under different wave types, the buoy's oscillating motion state will be significantly affected by specified wave types. The trajectories of water particles of traveling waves are closed ellipses. So, there are no differences for the corresponding trajectories of water particles on free surface at any position. By comparison, when the wave field of the flume is composed of standing waves and the oscillating buoy of WEC is arranged in the wave antinodes, because the water particles of standing waves oscillate vertically at the wave antinodes, the buoy is only oscillating vertically under the influence of the water particles. During the experiment, in order to study the effects of different types of waves on the performance of oscillating buoy WEC, traveling waves, fully standing waves and partial standing waves (induced by Bragg resonance) of the same wave height were generated in the wave flume, respectively. In other words, on the basis of the calibration stage, the wave heights in the travelling wave cases are kept identical to the wave height values at the antinodes in the corresponding partial and fully standing waves' cases. Furthermore, the generating power of the device was compared and analyzed.

The experiment was carried out in the flume of Estuary Waterway Experiment Hall of Hohai University, which is 67 m long, 1 m wide and 1.5 m deep (Figure 2). The front end of the flume was equipped with a plate-type wave maker, which can stably generate waves corresponding to various operating conditions required for the experiment. The back end of the flume was installed with the rubble slope, which can effectively dissipate the effects of wave reflection. Along the flume, 16 wave probes were arranged to measure the wave field in the flume primarily for the calibration of wave parameters and corresponding wave field status in flume.

Figure 2. Sketch of the experimental flume.

For wave energy conversion, a buoy was made by a hollow steel buoy of elliptical cylinder (Figure 3), which was installed at the incident side of the wavy bottoms with a distance of 33 m to the wave maker. Its dimensions are: the semi-major axis R_a = 0.75 m, semi-minor axis R_b = 0.3 m, height H_{buoy} = 0.45 m, quality m = 25 kg. The reason for choosing the buoy with an elliptical cross-section was that its minor axis occupies less space of the width of the flume; reflection of incident waves can be reduced to obtain a stabler wave field.

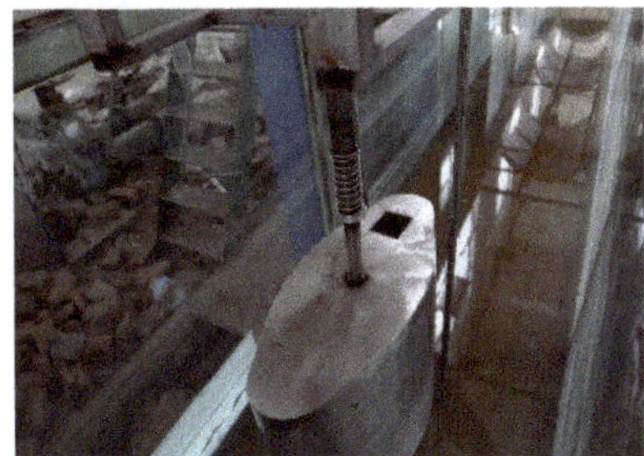

Figure 3. The elliptical cylinder of oscillating buoy.

The buoy adopted permanent magnet single phase linear motor which was arranged 33 m away from the end of the flume (Figure 4). The lead wires of the three disk windings of the primary part of the motor (stator) were connected successively in series. A spring structure was installed at the end of the secondary part of the motor (rotor) to adjust its motion length according to the actual wave height. The rotor was connected with an elliptical cylindrical buoy and the buoy floated on the water surface. The stator was fixed on the flume, and the primary winding was connected with the rectifier circuit (including the load bulb and electric power recorder). The detailed information of motor's performance has been stated by Yu et al. [37] and Huang et al. [38]. The voltage and current signals outputted from the motor were collected and recorded by the electric power recorder. The overall arrangement of the WEC experimental devices in this study is shown in Figure 5. It should be noted that only the heaving motion is allowed for the oscillating buoy for wave energy conversion. Furthermore, the spring structures at the up and downside of the cylinder are primarily applied to limit the extreme vertical displacement of the rotor.

(1) For the cases of partial standing waves, the corresponding wave field are induced by Bragg resonance. In the study of the effects of Bragg resonance on the performance of wave energy generation [35], five fixed continuous sinusoidal wavy bottoms (Figure 6) were installed in the flume to induce the Bragg resonance, whose wavelength is 0.9 m and height is 0.3 m. Figure 7 presents the experimental devices' layout for partial standing waves induced by Bragg resonance.

Figure 4. The permanent magnet single phase linear motor installed in flume.

Figure 5. Installation of WEC experimental device in flume.

Figure 6. Five fixed continuous sinusoidal wavy bottoms installed in flume.

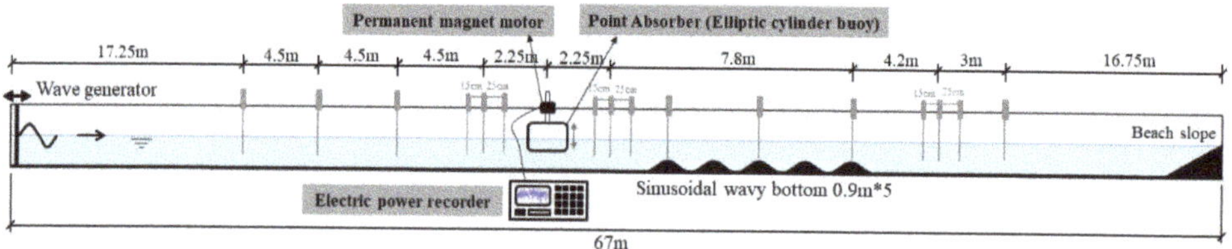

Figure 7. Experimental devices layout diagram for partial standing wave cases (Adapted from [35] with permission from Springer Nature).

According to the dispersion relation and the bottom designed, the incident wave period corresponding to the resonance peak T_p is 1.15 s, which means that the Bragg resonance effect is strongest at 1.15 s. Thus, the value of incident wave period T is centered on 1.15 s and varies within an appropriate range, in which case strong Bragg resonance can be generated. Based on the analysis above, the non-dimensional incident wave period

T/T_p of this experiment is 0.96, 1.0, 1.04, the water depth is 0.6 m, 0.7 m, 0.8 m and all the incident waves are regular. Meanwhile, the wave field on the incident side of the rippled bottoms is partial standing waves when Bragg resonance occurs. Thus, the measurement of the power generation of the device under partial standing waves conditions (caused by the Bragg resonance) has been completed. For all the cases of partial standing waves, the locations of the antinodes have been determined at the calibration stage. Then the oscillating buoy will be placed at the corresponding location of antinode for each case.

In this work, we additionally performed the flume experiment to measure the power generation of the device under traveling waves and fully standing waves and set the same wave parameters for comparative analysis.

(2) For the cases of travelling waves, the wave height of travelling waves at the location of buoy was calibrated to be the same as partial standing waves [35]. In the cases of travelling waves, all five rippled bottoms have been removed from the flume. The flume's bottom is completely flat in this situation (as shown in Figure 8). Furthermore, the oscillating buoy WEC was still installed at the same location.

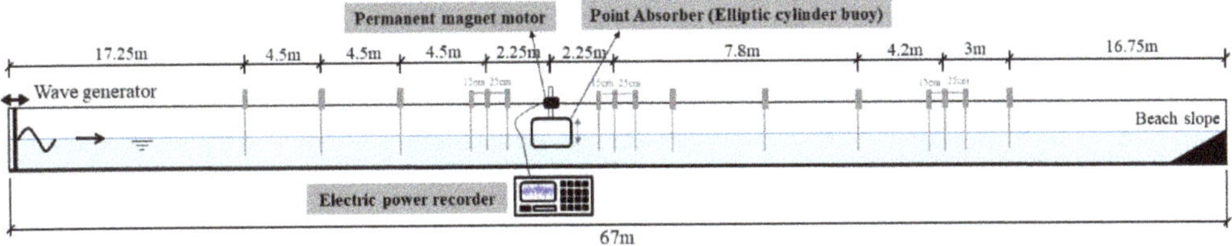

Figure 8. Experimental devices layout diagram for travelling wave cases.

(3) For the cases of fully standing waves, in order to create the wave field of fully standing waves, a vertical rigid baffle was installed in the end of flume to reflect the incident waves completely (Figure 9). Furthermore, wave height of fully standing waves at the location of buoy was set and adjusted to be same as traveling waves and partial standing waves (Figure 10).

Figure 9. Wooden rigid vertical baffle (inducing fully standing waves) installed in flume.

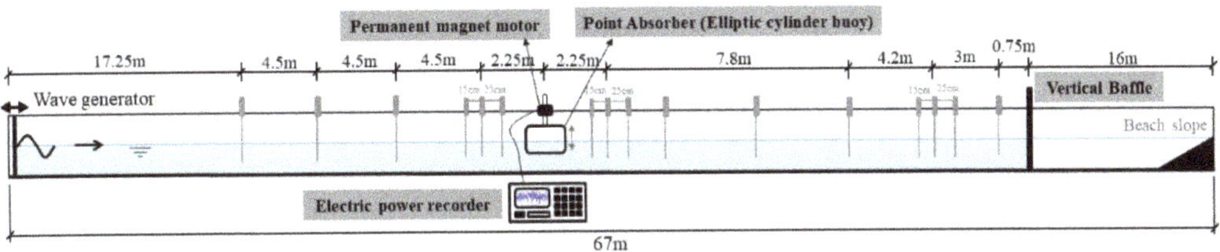

Figure 10. Experimental devices layout diagram for fully standing wave cases.

On the basis of this, the buoy was installed after the wave height calibration, and the power of the oscillating buoy WEC was measured. The vertical baffle was set at 0.75 m away from the last wave probe in order to keep enough space for reflected waves (by rigid vertical baffle firstly and wave paddle secondly) to avoid disturbing the wave field around buoy. So, there will be sufficient time to keep stable standing wave region near buoy (avoid transverse sloshing in the wave field) and to measure the power of the WEC. Similar to the cases of partial standing waves, the oscillating buoys were always located at the antinodes which have been calibrated and determined. The experimental cases and conditions are listed in Table 1.

Table 1. The experiment group arrangements.

Wave types	Cases	Water Depth h (m)	Wave Period T (s)	T/T_p	Wave Height at the Buoy H (m)	R_{Bragg}
Traveling waves/ Fully standing waves/ Partial standing waves (by Bragg resonance) Note: Nine cases applied to all three wave types correspondingly.	1	0.6	1.1	0.96	0.109	0.3915
	2	0.6	1.15	1.0	0.119	0.5219
	3	0.6	1.2	1.04	0.118	0.5173
	4	0.7	1.1	0.96	0.109	0.2929
	5	0.7	1.15	1.0	0.119	0.3268
	6	0.7	1.2	1.04	0.118	0.2764
	7	0.8	1.1	0.96	0.109	0.1783
	8	0.8	1.15	1.0	0.119	0.1955
	9	0.8	1.2	1.04	0.118	0.1296

Note: R_{Bragg} and T/T_p only applied to the situation of partial standing waves induced by Bragg resonance. R_{Bragg} is the Bragg resonance's reflection coefficient measured at the calibration stage before the installation of oscillating buoy (only applied to the situation of partial standing waves). T_p is the period of resonance peak in the cases above at three different water depths.

It should be noted that the minimum water depth set in our flume experiment is 0.6 m. Although the smaller water depth will induce the stronger Bragg resonance, the reflected wave component will break above the rippled bottoms if the water depth h is less than 0.6 m. This is because the water depth above the bottoms' crest will be (h-H_b) which is much smaller. Therefore, the lower bound value of the water depth in our experiment is set as 0.6 m. Besides, the electrical signals measured will be discussed in the next section without the free-surface elevation measured in flume. The reason is that the wave fields were interfered by oscillating buoy (irregular reflection) during the experimental process. Furthermore, the wave patterns observed at the experimental stage are not considerable to be compared with the electric signals observed by electric power recorder. So, the free-surface elevations measured by wave probes along the flume are utilized for calibrating the wave parameters and wave field spatial distribution in this study.

During the calibration stage of flume experiment, the wave height and period are calculated by the zero-upcrossing method. Furthermore, the reflection coefficient is calculated by the incident-reflected separation method of Mansard [39]. The detailed information of the reflection coefficient by Bragg resonance has been introduced in Tao [35].

4. Results and Discussions

In this section, we discuss the electrical energy generation characteristics of traveling waves, partial standing waves and fully standing waves by instantaneous voltage, instantaneous current, instantaneous power and active power respectively. In addition, the electrical energy generation performance will be compared by average active power in detail for three different wave types.

4.1. Comparison of Measured Instantaneous Voltage and Current

The electric power recorder is utilized to record the power generation of the oscillating buoy WEC under each case and to collect the instantaneous current and voltage data. The sampling frequency of the electric power recorder is 1000 Hz, and the data acquisition starts from the wave propagation to the buoy and the duration is 1 min.

Figures 11–13 show the measured voltage time series diagrams when the wave types are traveling waves, partial standing waves and fully standing waves, respectively. The water depth $h = 0.6$ m and the incident wave period $T/T_p = 1.0$. From Figure 11, it could be easily found that, due to the wave front induced at the initial stage of wave generation by wave paddle, the voltage amplitude also has the similar pattern around 1 s of its time series. Then the incident wave at the buoy begins to be stable and regular, and the voltage waveform tends to be stable correspondingly. In Figure 12, the initial waveform also fluctuates because of the effect of the wave group. Later, due to the superposition of incident waves and reflected waves caused by Bragg resonance, partial standing waves are formed in flume, and the wave peak of the instantaneous voltage increases gradually. The stable state is reached in about 20 s, which coincides with the calculated time when the reflected wave induced by Bragg scattering propagates to the buoy. Compared with Figure 11, when the wave field is stabilized under the condition of partial standing waves, the measured instantaneous voltage peak is larger than that under traveling waves. Figure 13 gives the instantaneous voltage waveform under fully standing waves. After the buoy motion is stable, the voltage peak is close to that in partial standing waves.

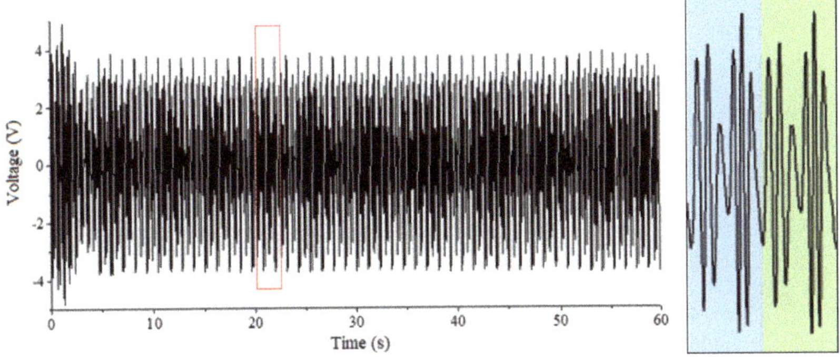

Figure 11. Voltage diagram (including the zoom-up of the enclosed region between 20.0 s and 22.4 s) under traveling waves of $T/T_p = 1.0$, $h = 0.6$ m.

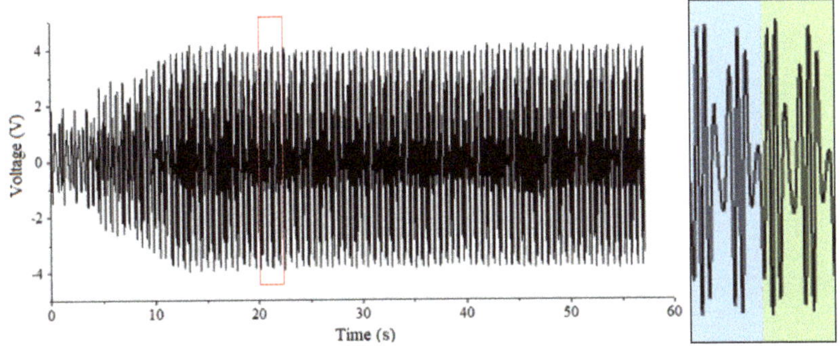

Figure 12. Voltage diagram (including the zoom-up of the enclosed region between 20.0 s and 22.3 s) under partial standing waves of $T/T_p = 1.0$, $h = 0.6$ m (Reproduced from [35] with permission from Springer Nature).

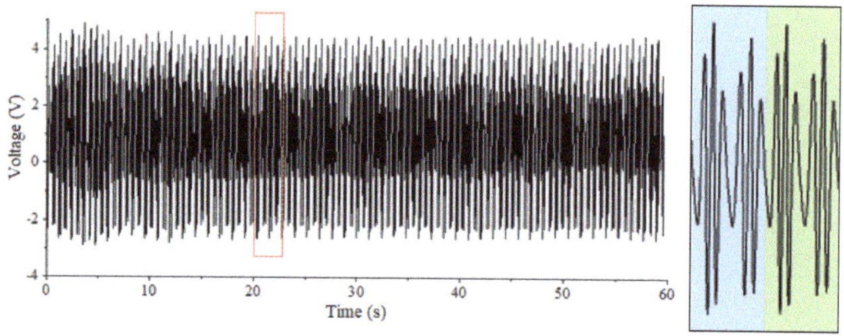

Figure 13. Voltage diagram (including the zoom-up of the enclosed region between 20.1 s and 22.5 s) under fully standing waves of $T/T_p = 1.0$, $h = 0.6$ m.

Figures 14–16 show the current time series diagrams measured under the conditions that the wave types are traveling waves, partial standing waves and fully standing waves, respectively. The trends in current variation time series could not be identified obviously in these three charts. After the wave field is stabilized in the flume, the peak value of instantaneous current measured under fully standing waves is the largest, second for partial standing waves and the smallest for traveling waves.

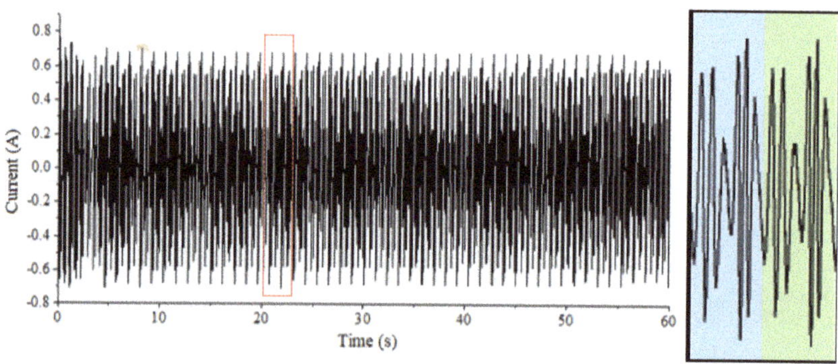

Figure 14. Current diagram (including the zoom-up of the enclosed region between 20.0 s and 22.4 s) under traveling waves of $T/T_p = 1.0$, $h = 0.6$ m.

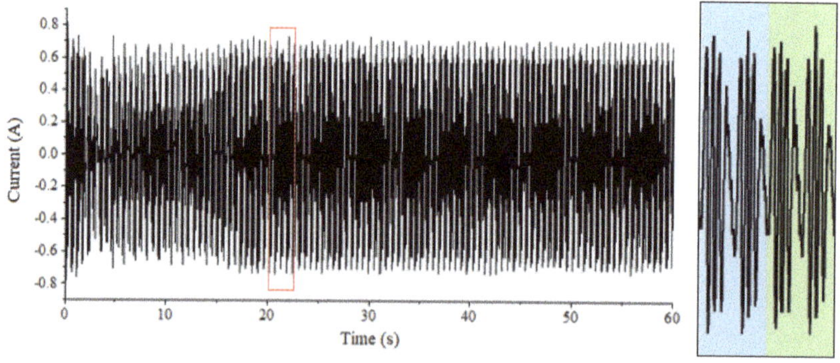

Figure 15. Current diagram (including the zoom-up of the enclosed region between 20.0 s and 22.3 s) under partial standing waves of $T/T_p = 1.0$, $h = 0.6$ m.

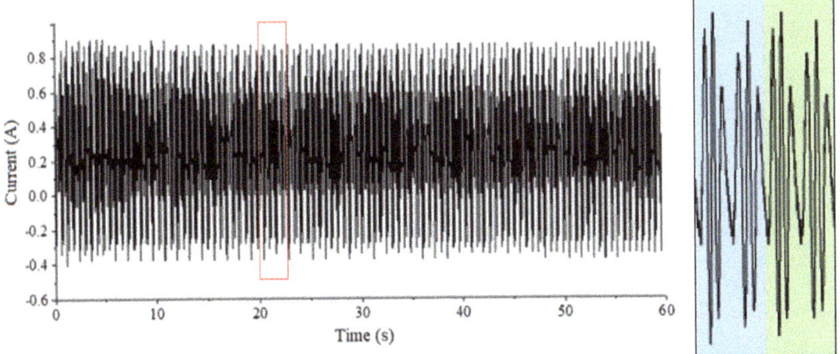

Figure 16. Current diagram (including the zoom-up of the enclosed region between 20.1 s and 22.5 s) under fully standing waves of $T/T_p = 1.0$, $h = 0.6$ m.

4.2. Comparison of the Instantaneous Power

The instantaneous power can reflect the instantaneous absorbed power of the circuit, which is helpful to analyze the power variation of the WEC with the change of the wave field. Taking the incident wave period $T/T_p = 1.0$ and the water depth of 0.6 m as an example, the instantaneous power of traveling waves, partial standing waves and fully standing waves is compared and analyzed, respectively.

The electric power recorder begins to collect data since the arrival of the incident waves to the buoy, and the acquisition time is 1 min. It should be noted that the influence of irregular wave packets (wave front) appeared at the very beginning of wave recording, a stable standing wave region has not yet been formed. In addition, the reflected waves generated by the incident waves propagating to the end of the flume will affect the experimental results. Thus, when calculating the power generation, the data used should be within the interval from the arrival of Bragg reflection waves at the buoy to the arrival of the reflected waves at the buoy.

The waveform propagation velocity (phase velocity) c is given by

$$c = \frac{L}{T} \tag{8}$$

in which L is the wavelength and T is the corresponding wave period.

The wave energy propagation velocity c_g is given by

$$c_g = cn = \frac{c}{2}\left(1 + \frac{2kh}{\sinh 2kh}\right) \tag{9}$$

in which, k is the wavenumber and h is the water depth.

The velocity of wave energy propagation in Equation (9) is utilized to calculate the propagation time of wave train in order to estimate the effective range in total time series of measurements. Furthermore, the effective time t_e can be expressed as:

$$t_e = \frac{s}{c_g} \tag{10}$$

in which, s is the length wave energy propagating path. The lower bound of s represents the propagating length of initially induced wave train. The upper bound of s usually depends on the total propagating length of disturbance wave components before reaching the measurement location (i.e., reflected waves from wave maker or bottom slope).

Taking the case of period $T/T_p = 1.0$, water depth $h = 0.6$ m as an example, the actual wave energy propagation velocity is 1.0035 m/s. The wave recording begins from the steady waveform to the buoy, and the Bragg reflection waves arrive at the buoy after

20.03 s. After 72.25 s, the reflected waves reach the buoy. The recording time is 1 min, so the range used to calculate the power is 20.03 s~60 s. When calculating the instantaneous power of the WEC under fully standing waves, due to the change of the experiment model and corresponding set-up, the wave recording begins after 15 s from the steady wave to the buoy, and completely standing waves reach the buoy after 20.87 s. The range of power available for calculation is 20.87 s~60 s. In general, this time interval analysis aims at selecting the time series which are effective to capture the Bragg reflection (or baffle reflection) waves and avoid the reflected interference waves by both sides of the flume (wave paddle and/or rubble mounds).

Thus, the instantaneous power variation of WEC within 25 s~35 s under various wave types is compared. The instantaneous power of traveling waves, partial standing waves and fully standing waves is calculated respectively.

The formula for calculating instantaneous power p is as follows:

$$p = ui \tag{11}$$

Here u denotes the voltage, and i denotes the current.

Figures 17–19 show the comparative diagram of instantaneous power of the WEC under the conditions of traveling waves, partial standing waves and fully standing waves respectively when the water depth $h = 0.6$ m and the incident wave period $T/T_p = 1.0$. It can be seen from these three figures that the instantaneous power peak of the WEC under traveling waves is stable at about 2.5 W. Furthermore, the peak value under partial standing waves is larger than 3 W. The instantaneous power peak of the WEC under fully standing waves exceeds 3.5 W, but its waveform is more irregular than that of traveling waves and partial standing waves, which indicates that the motion state of buoy is unstable during power generation. When the incident wave period $T/T_p = 1.0$, a total of 17 peaks occur within 10 s, and 8.7 waves are measured in the same time period. The calculated peak number of generations is exactly twice the number of waves.

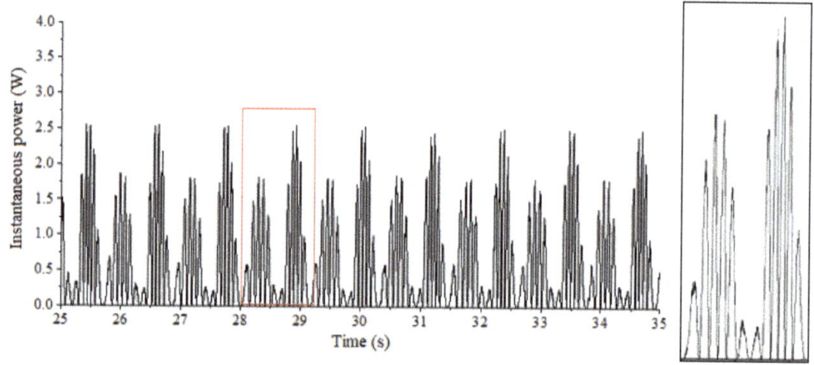

Figure 17. Instantaneous power diagram (including the zoom-up of the enclosed region between 28.0 s and 29.2 s) under traveling waves of $T/T_p = 1.0$, $h = 0.6$ m.

4.3. Comparison of the Active Power

As in the calculation of instantaneous power, the active power of the oscillating buoy WEC under different wave types is also calculated. The related calculation method and expressions could be referred by Equations (3) and (4) in the previous research of Tao et al. [35]. Furthermore, the obtained results are shown in Figures 20–22, respectively, which are under conditions of traveling waves, partial standing waves and fully standing waves. Data from 22 s to 37 s are selected to calculate the active power. Active power, also known as the average power, is the average value of the instantaneous power integral consumed by the load within a period.

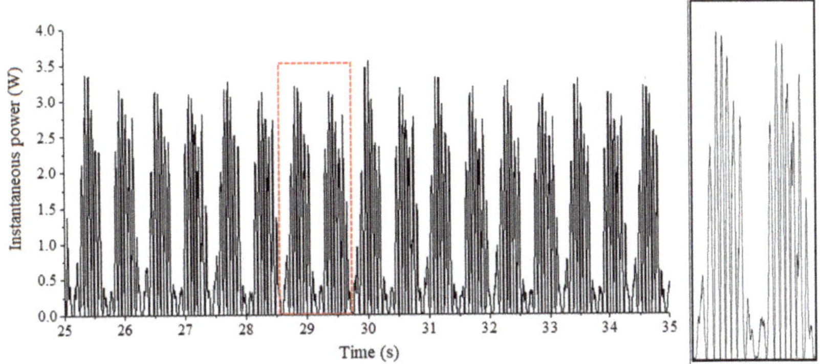

Figure 18. Instantaneous power diagram (including the zoom-up of the enclosed region between 28.6 s and 29.8 s) under partial standing waves of $T/T_p = 1.0$, $h = 0.6$ m.

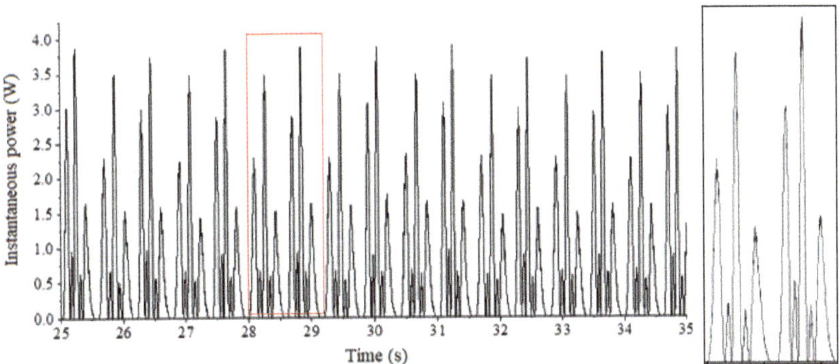

Figure 19. Instantaneous power diagram (including the zoom-up of the enclosed region between 28.0 s and 29.1 s) under fully standing waves of $T/T_p = 1.0$, $h = 0.6$ m.

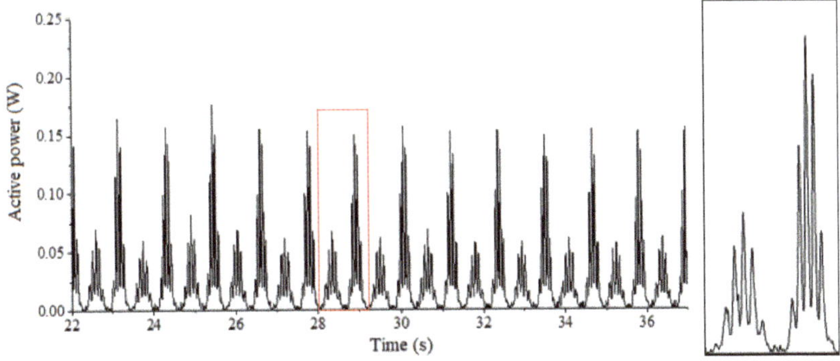

Figure 20. Active power diagram (including the zoom-up of the enclosed region between 28.0 s and 29.2 s) under traveling waves of $T/T_p = 1.0$, $h = 0.6$ m.

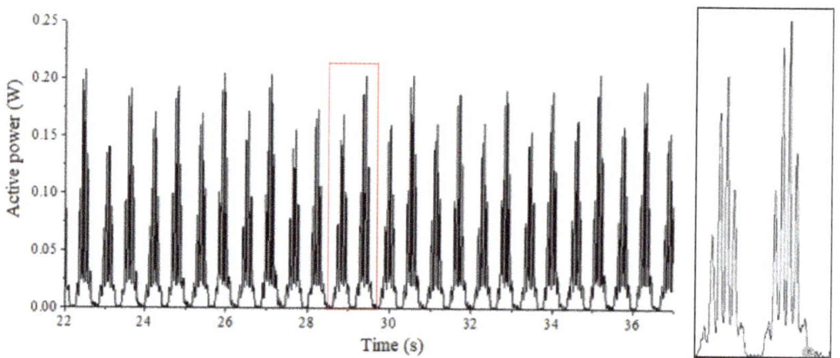

Figure 21. Active power diagram (including the zoom-up of the enclosed region between 28.6 s and 29.8 s) under partial standing waves of $T/T_p = 1.0$, $h = 0.6$ m.

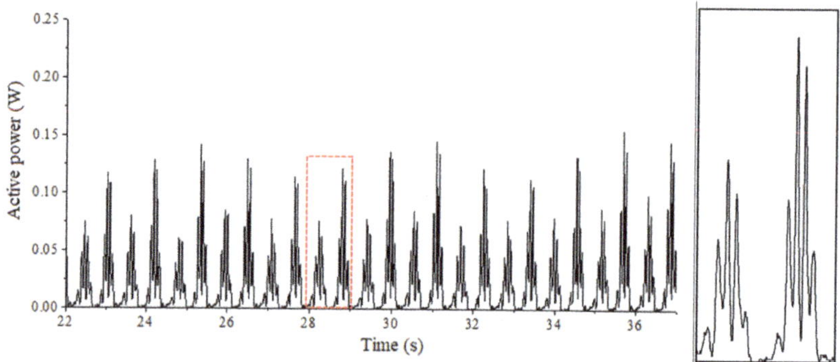

Figure 22. Active power diagram (including the zoom-up of the enclosed region between 28.0 s and 29.1 s) under fully standing waves of $T/T_p = 1.0$, $h = 0.6$ m.

By comparison, when the water depth $h = 0.6$ m, the period $T = 1.5$ s and the wave type is partial standing waves, the active power value of the oscillating buoy WEC is the maximum, and the amplitude reaches 0.2 W; when the wave type is traveling waves, the active power amplitude of the generator exceeds 0.16 W; when the waves in the flume are fully standing waves, the active power amplitude of the WEC is about 0.12 W. Within 15 s, there are 26 active power peaks and 13 waves in the same time period, which is the same as the previous calculation of instantaneous power, and the peak number of the device generation is exactly twice the number of waves.

The total wave energy per unit wave crest width over one wavelength of three wave types is analyzed from the perspective of energy. Assuming that the maximum wave height is 2H, when the wave type is traveling waves, the incident wave height is 2H. The total wave energy per unit wave crest width over one wavelength E_t is given on the basis of the expression from [36]:

$$E_t = \frac{1}{8}\rho g (2H)^2 = \frac{1}{2}\rho g H^2 \tag{12}$$

For fully standing waves, the incident and reflected wave height are H, and the total wave energy per unit wave crest width over one wavelength E_f is as follows:

$$E_f = \frac{1}{8}\rho g H^2 \times 2 = \frac{1}{4}\rho g H^2 \tag{13}$$

The total wave energy E_p per unit wave crest width over one wavelength of partial standing waves is related to its reflection coefficient, which is between fully standing waves and partial standing waves. Therefore, when the maximum wave height is the same, the total wave energy per unit wave crest width over one wavelength is compared as follows:

$$E_t > E_p > E_f \tag{14}$$

Through the comparison of active power, in the case of $T/T_p = 1.0$ and $h = 0.6$ m, it is found that the active power of WEC is the smallest under fully standing waves, slightly larger than that under traveling waves and the maximum under partial standing waves. This is different from the previous analysis of three types of wave energy. The reasons might be concluded twofold as follows:

Firstly, the performance of oscillating buoy WEC under different wave types is related to the ratio of buoy diameter to wavelength R/L. In traveling waves, the wave energy density is spatially uniform within a wavelength range. In fully standing waves, the wave energy density at the wave antinodes of a wavelength range is the largest, the wave energy density gradually decreases to the minimum at the wave nodes and it reaches zero at the wave nodes. Thus, in fully standing waves, the larger the ratio of buoy diameter to wavelength R/L, the lower the performance of WEC. While in partial standing waves with the same maximum wave height, the wave energy density is the same at the wave antinodes and decreases gradually to the minimum at the wave nodes, but the wave energy density at the wave nodes is not zero. Therefore, for large ratio values of R/L, the buoy of the same diameter has the lowest generation performance in fully standing waves. For the cases of $T/T_p = 1.0$, the values of R_a/L are 0.380, 0.373 and 0.368 for 0.6 m, 0.7 m and 0.8 m water depths correspondingly in this study.

Then, for comparison between the situations under travelling waves and partial standing waves, the performance of oscillating buoy WEC is also related to the trajectories of water particles. The trajectories of the particles of traveling waves are elliptical. Under the action of wave force, the buoy will have heaving, surging and rolling movements in water simultaneously. For the linear motor used in this paper, the heaving energy can be converted directly into electric energy, while the friction damping of the motor is increased by surging and rolling, which will significantly affect the performance of power generation. The trajectories of water particles motion of the partial standing waves are oscillating in the vertical direction, so when the center of gravity of the buoy is the same as the x-axis coordinate of the point of the wave antinodes, the buoy is mainly subjected to the vertical wave force and oscillating vertically, which weakens the surging and rolling motion of the buoy. Thus, the effect of friction damping on the performance of motor generation is reduced. Therefore, when the maximum wave height in the flume is the same, the energy conversion performance of partial standing waves is higher than that of traveling waves. So, in general, considering the combined effects of the water particles' trajectories and the ratio of buoy diameter to wavelength, the WEC could have the largest active power in partial standing waves.

4.4. Effects of Wave Types on the Performance of WEC under Different Wave Parameters

According to the active power calculation method, the active power curves corresponding to each group of working conditions in Table 1 are obtained. By summing the active power P in this time period, the average active power value $\bar{p}$ in that time period is obtained.

Due to the different velocity of wave energy propagation, the time period selected for calculation under different incident wave periods is not exactly the same. It is required that the active power waveform in the selected time period is more stable to ensure that the buoy is in a stable state. Under the same incident wave period, the calculated time period should be the same. If the incident wave periods are different, the starting point and the ending point of the calculated time period may be different correspondingly, but the length of the calculated time period keeps the same.

Figures 23–25 show the performance comparisons of oscillating buoy WEC under three different wave types. Among them, under the condition of partial standing waves, when the incident wave period $T/T_p = 1.0$, the average active power is the largest corresponding to the peak of Bragg reflection. In Figure 23, when the incident wave period $T/T_p = 0.96$, since the Bragg resonance effect is relatively weak and the standing wave height is not significantly increased with respect to the incident wave height, the average active power value of the WEC under three wave types is basically identical. When the incident wave period $T/T_p = 1.04$, the average active power of the WEC under partial standing waves is the highest, followed by that under traveling waves. Furthermore, the average active power under fully standing waves is the lowest. The reason might be that, when the maximum wave height is the same, the total wave energy per unit wave crest width over one wavelength of the three different wave types is different, and the relationship can be expressed as $E_t > E_p > E_f$. However, the performance of wave energy generation is not only related to the density of wave energy but also to the motion trajectories of water particles. The buoy's motion not only includes heaving but also surging and rolling under the condition of traveling waves, while the buoy only has heaving motion in partial standing waves at the wave antinodes. So, for the linear motor utilized in our flume experiments, the friction damping of the buoy decreases under the condition of partial standing waves, thus increasing the energy conversion performance. Therefore, the final total performance of the WEC under partial standing waves is higher than that in traveling waves.

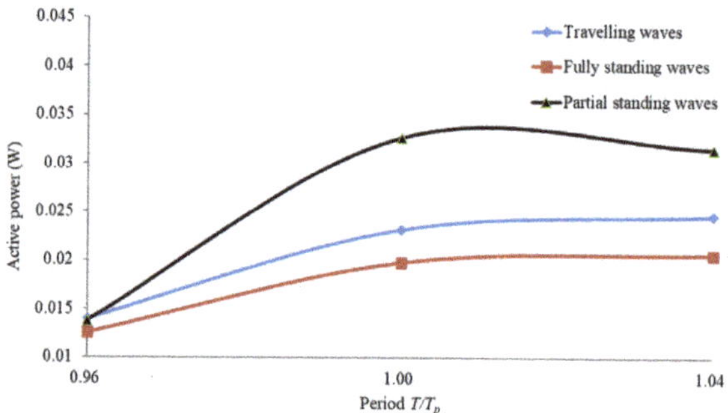

Figure 23. Comparison of generation performance under three different wave types at $h = 0.6$ m.

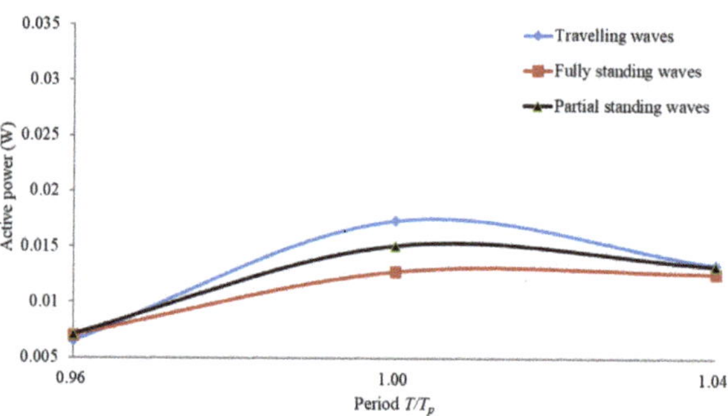

Figure 24. Comparison of generation performance under three different wave types at $h = 0.7$ m.

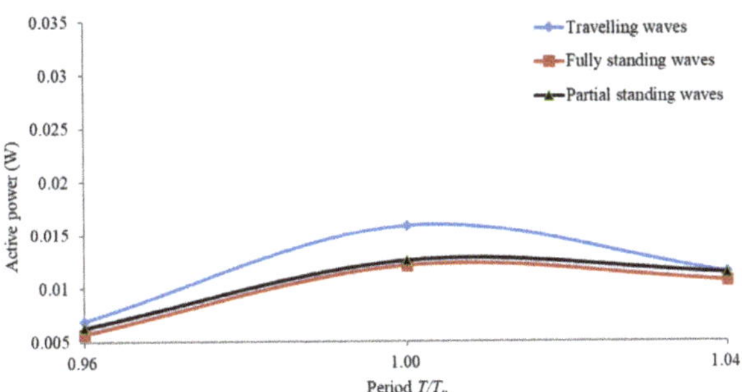

Figure 25. Comparison of generation performance under three different wave types at $h = 0.8$ m.

Figures 24 and 25 compare the performance of wave power generation device under three different wave types when the water depth of the flume is 0.7 m and 0.8 m, respectively. In these two figures, the oscillating buoy WEC has the highest power generation performance in traveling waves, which is second in partial standing waves and lowest in fully standing waves. In Figure 25, the difference between the total wave energy and the energy conversion performance of these two kinds of (partial and fully) standing waves reduces further, so the generation performance is similar at different periods' conditions. This might be related to the effects of increasing wavelength at larger water depth (compared with same buoy's size, especially to the behavior of buoy's oscillation under partial and fully standing wave conditions), which is stated in detail after.

Overall, when the water depth $h = 0.6$ m, the generation performance of partial standing waves is 39% higher than that of traveling waves and 65% higher than that of fully standing waves. For this condition (with large wave steepness and small wavelength), the improvement of the energy conversion performance of WEC by (partial) standing wave types along with the effects of associated water particles' motion is quite obvious.

For the cases with larger water depth, the lifting effect of standing waves on the oscillating buoy wave power generation device is limited due to the smaller difference of the wave energy density and the decrease of the vertical motion distance of the buoy. In the following, the effects of ratio of buoy's semi-major axis to the wavelength will be discussed. Under the condition of relatively large ratio of the semi-major axis to the wavelength of buoy R_a/L (small L with small water depth) in this experiment, the differences of effects between fully standing waves and partial standing waves on the performance of generator are prominent at the same wave height. When the value of the R_a/L is small, the buoy could be nearly regarded as a particle moving with free-surface oscillation, and the wave energy density within the buoy's wave energy length range is basically identical. At this time, the generation performance of partial standing waves is the same as that of fully standing waves. When the value of the R_a/L is large, the generation power of fully standing waves at the same wave height is smaller than that of partial standing waves and traveling waves, which is worth noting. This might be related to more complex nonlinear behaviors of water particles' motion and buoy's response, which need further investigation in the future. Besides, in China, some representative wave periods values are 3.2 s, 3.8 s, 4.0 s and 4.4 s, and the corresponding deep-water wavelengths are 16.0 m, 22.5 m, 25.0 m and 30.2 m respectively. For rough estimation of the effect of buoy's size, their estimated ratios of buoy's size to wavelength are no longer small values. Therefore, if the wave types' influence is considered to improve the performance of oscillating buoy WEC, the effect of buoy size on generation performance should be evaluated (to avoid over-sized buoy device).

5. Conclusions

In this work, the influence of different wave types (travelling waves, partial and fully standing waves) on the energy conversion performance of oscillating buoy WEC is preliminarily studied by flume experiments. The measured instantaneous voltage, instantaneous current, instantaneous power and active power of the WEC are analyzed and compared under three different wave types. It is found that standing wave field could enhance the energy conversion performance of the oscillating buoy WEC to some extent. By representative experimental cases with 0.6 m water depth and the incident wave period $T/T_p = 1.0$, the performance of the WEC in partial standing waves is increased by 39% with respect to traveling waves of the same wave height, and this increment could not be ignored if the multiple submerged breakwaters exist in the seabed in nearshore areas.

Although the wave energy density among these three wave types (largest in traveling waves, second in partial standing waves and minimal in fully standing waves) is not consistent with the measurements concluded above, the trajectories could have more connections to the kinetical and dynamical behavior of oscillating buoy WEC facilities. Besides, higher wave steepness will increase the vertical motion distance of the buoy to enlarge the effects of standing waves on the performance of wave energy conversion. It should also be noted that, from the perspective of relative size of the buoy compared with the wavelength, the performance of WEC is also related to the ratio of buoy semi-major axis to wavelength R_a/L . Due to the relatively large ratio R_a/L in our flume experiment, the generation power of fully standing waves is smaller than that of partial standing waves and traveling waves. It reveals that R_a/L should be limited when standing wave field is utilized to improve generation performance. So, the comparison results of dynamic response of oscillating buoy between fully and partial standing waves in this work might be limited with our experimental set-up. Furthermore, the detailed connections between oscillating buoy's behavior and R_a/L under fully and partial standing waves will be perused underway. Besides, the methodology in this study is based on flume experiments. In order to obtain more detailed response of the WEC oscillating buoy devices, the numerical simulation as well as its comparison with experimental results is necessary as future work.

Author Contributions: Data curation, Y.W.; Formal analysis, S.Q. and H.Z.; Investigation, J.F. and Y.W.; Methodology, S.Q. and J.S.; Software, J.S.; Validation, H.Z.; Writing–original draft, S.Q.; Writing–review & editing, J.F. All authors have read and agreed to the published version of the manuscript.

Funding: This research work was financially supported by the National Natural Science Foundation of China (Grant Nos. 51579091, U1706230).

Institutional Review Board Statement: Not applicable.

Informed Consent Statement: Not applicable.

Data Availability Statement: Not applicable.

Conflicts of Interest: The authors declare no conflict of interest.

References

1. Salter, S.H. World progress in wave energy. *Int. J. Ambient. Energy* **1989**, *10*, 3–24. [CrossRef]
2. Falnes, J. A review of wave-energy extraction. *Mar. Struct.* **2007**, *20*, 185–201. [CrossRef]
3. Alain, C.; Mccullen, P.; Falcão, A.; Fiorentino, A.; Gardner, F.; Hammarlund, K.; Lemonis, G.; Lewis, T.; Nielsen, K.; Petroncini, S.; et al. Wave energy in Europe: Current status and perspectives. *Renew. Sustain. Energy Rev.* **2002**, *6*, 405–431.
4. Falcão, A.F.O. Wave energy utilization: A review of technologies. *Renew. Sustain. Energy Rev.* **2010**, *14*, 899–918. [CrossRef]
5. Czech, B.; Bauer, P. Wave Energy Converter Concepts: Design Challenges and Classification. *IEEE Ind. Electron. Mag.* **2012**, *6*, 4–16. [CrossRef]
6. Chandrasekaran, S.; Raghavi, B. Design, Development and Experimentation of Deep Ocean Wave Energy Converter System. *Energy Procedia* **2015**, *79*, 634–640. [CrossRef]
7. Malmo, O. Wave-power absorption by an oscillating water column in a reflecting wall. *Appl. Ocean Res.* **1986**, *8*, 42–48. [CrossRef]
8. Rahm, M.; Svensson, O.; Boström, C.; Waters, R.; Leijon, M. Experimental results from the operation of aggregated wave energy converters. *IET Renew. Power Gener.* **2012**, *6*, 149–160. [CrossRef]
9. López, I.; Andreu, J.; Ceballos, S.; De Alegría, I.M.; Kortabarria, I. Review of wave energy technologies and the necessary power-equipment. *Renew. Sustain. Energy Rev.* **2013**, *27*, 413–434. [CrossRef]
10. Mei, C.C. Hydrodynamic principles of wave power extraction. *Philos. Trans. A Math. Phys Eng. Sci.* **2012**, *370*, 208–234. [CrossRef]

11. Lovas, S.; Mei, C.C.; Liu, Y. Oscillating water column at a coastal corner for wave power extraction. *Appl. Ocean Res.* **2010**, *32*, 267–283. [CrossRef]
12. Luo, Y.; Nader, J.R.; Cooper, P.; Zhu, S.P. Nonlinear 2D analysis of the efficiency of fixed Oscillating Water Column wave energy converters. *Renew. Energy* **2014**, *64*, 255–265. [CrossRef]
13. Drew, B.; Plummer, A.; Sahinkaya, M.N. A review of wave energy converter technology. *J. Power Energy* **2009**, *223*, 887–902. [CrossRef]
14. Zheng, S.; Zhang, Y. Theoretical modelling of a new hybrid wave energy converter in regular waves. *Renew. Energy* **2018**, *128*, 125–141. [CrossRef]
15. Serman, D.D.; Mei, C.C. Note on Salter's energy absorber in random waves. *Ocean Eng.* **1980**, *7*, 477–490. [CrossRef]
16. Ning, D.Z.; Zhou, Y.; Mayon, R.; Johanning, L. Experimental investigation on the hydrodynamic performance of a cylindrical dual-chamber Oscillating Water Column device. *Appl. Energy* **2020**, *260*, 114252. [CrossRef]
17. Poguluri, S.K.; Kim, D.; Bae, Y.H. Hydrodynamic Analysis of a Multibody Wave Energy Converter in Regular Waves. *Processes* **2021**, *9*, 1233. [CrossRef]
18. Poguluri, S.K.; Ko, H.S.; Bae, Y.H. CFD investigation of pitch-type wave energy converter-rotor based on RANS simulations. *Ships Offshore Struct.* **2020**, *15*, 1107–1119. [CrossRef]
19. Korde, U.A. Systems of reactively loaded coupled oscillating bodies in wave energy conversion. *Appl. Ocean Res.* **2003**, *25*, 79–91. [CrossRef]
20. Tom, N.M.; Madhi, F.; Yeung, R.W. Power-to-load balancing for asymmetric heave wave energy converters with nonideal power take-off. *Renew. Energy* **2017**, *131*, 1208–1225. [CrossRef]
21. Shi, H.; Cao, F.; Liu, Z.; Qu, N. Theoretical study on the power take-off estimation of heaving buoy wave energy converter. *Renew. Energy* **2016**, *86*, 441–448. [CrossRef]
22. Malmo, O.; Reitan, A. Wave-power absorption by an oscillating water column in a channel. *J. Fluid Mech.* **2006**, *158*, 153–175. [CrossRef]
23. Vantorre, M.; Banasiak, R.; Verhoeven, R. Modelling of hydraulic performance and wave energy extraction by a point absorber in heave. *Appl. Ocean Res.* **2004**, *26*, 61–72. [CrossRef]
24. Wu, F.; Zhang, X.P.; Ju, P.; Sterling, M.J. Modeling and Control of AWS-Based Wave Energy Conversion System Integrated into Power Grid. *IEEE Trans. Power Syst.* **2008**, *23*, 1196–1204.
25. Valerio, D.; Beirao, P.; da Costa, J.S. Optimization of wave energy extraction with the Archimedes wave swing. *Ocean Eng.* **2007**, *34*, 2330–2344. [CrossRef]
26. Shi, H.; Liu, Z.; Gao, R. Numerical investigation on combined oscillating body wave energy convertor. In Proceedings of the 2012 Oceans—Yeosu, Yeosu, Korea, 21–24 May 2012; IEEE: Piscataway, NJ, USA, 2012; pp. 1–5.
27. Cheng, B. *Design and Research of Wave Power Generation Device*; Shandong University: Jinan, China, 2012.
28. Gou, Y.F.; Ye, J.Q.; Li, F.; Wang, D. Investigation on the wave power device. *Acta Energ. Sol. Sin.* **2008**, *4*, 498–501.
29. Nielsen, S.R.K.; Zhou, Q.; Basu, B.; Sichani, M.T.; Kramer, M.M. Optimal control of an array of non-linear wave energy point converters. *Ocean Eng.* **2014**, *88*, 242–254. [CrossRef]
30. Tokić, G.; Yue, D.K.P. Optimal Configuration of Large Arrays of Floating Bodies for Ocean Wave Energy Extraction. Ph.D. Thesis, Massachusetts Institute of Technology, Department of Mechanical Engineering, Cambridge, MA, USA, 2016.
31. Folley, M.; Whittaker, T.J.T. The effect of sub-optimal control and the spectral wave climate on the performance of wave energy converter arrays. *Appl. Ocean Res.* **2009**, *31*, 260–266. [CrossRef]
32. Do, H.T.; Dang, T.D.; Ahn, K.K. A multi-point-absorber wave-energy converter for the stabilization of output power. *Ocean Eng.* **2018**, *161*, 337–349. [CrossRef]
33. Amini, E.; Golbaz, D.; Amini, F.; Majidi Nezhad, M.; Neshat, M.; Astiaso Garcia, D. A parametric study of wave energy converter layouts in real wave models. *Energies* **2020**, *13*, 6095. [CrossRef]
34. Neshat, M.; Sergiienko, N.Y.; Amini, E.; Majidi Nezhad, M.; Astiaso Garcia, D.; Alexander, B.; Wagner, M. A New Bi-Level Optimisation Framework for Optimising a Multi-Mode Wave Energy Converter Design: A Case Study for the Marettimo Island, Mediterranean Sea. *Energies* **2020**, *13*, 5498. [CrossRef]
35. Tao, A.F.; Yan, J.; Wang, Y.; Zheng, J.H.; Fan, J.; Qin, C. Wave power focusing due to the Bragg resonance. *China Ocean Eng.* **2017**, *31*, 458–465. [CrossRef]
36. Dean, R.G.; Dalrymple, R.A. *Water Wave Mechanics for Engineers and Scientists*; World Scientific Publishing Company: Singapore, 1991; pp. 79–100.
37. Yu, H.; Liu, C.; Yuan, B.; Hu, M.; Huang, L.; Zhou, S. A permanent magnet tubular linear generator for wave energy conversion. *J. Appl. Phys.* **2012**, *111*, 07A741. [CrossRef]
38. Huang, L.; Yu, H.; Hu, M.; Liu, C.; Yuan, B. Research on a tubular primary permanent-magnet linear generator for wave energy conversions. *IEEE Trans. Magn.* **2013**, *49*, 1917–1920. [CrossRef]
39. Mansard, E.; Funke, E.R. The measurement of incident and reflected spectra using a least squares method. *Coast. Eng. Proc.* **1980**, *1*, 154–172.

MDPI

St. Alban-Anlage 66

4052 Basel

Switzerland

Tel. +41 61 683 77 34

Fax +41 61 302 89 18

www.mdpi.com

Journal of Marine Science and Engineering Editorial Office

E-mail: jmse@mdpi.com

www.mdpi.com/journal/jmse

www.ingramcontent.com/pod-product-compliance
Lightning Source LLC
LaVergne TN
LVHW070914170726
843515LV00005B/777